SMITH, John Maynard. On evolution. Edinburgh (dist. by Aldine), 1972. 125p il tab 72-77391. 5.50. ISBN 0-85224-229-8

A selection of papers on evolutionary theory written by the author. Most were originally published in journals and symposia, although one was written for this book. Smith, an eminent British biologist and writer, deals with the status of evolutionary theory today. He points out that if the neo-Darwinian view is to be discredited, the most effective way of doing so would be to discredit its underlying Weismannist assumption of the noninheritance of acquired characters. If this happens, it will come from biochemical genetics, for in molecular terms the Weismannist assumption states that information flows from nucleic acids to protein and not in reverse. Smith concludes that it is unlikely that wholly new evolutionary processes remain to be discovered for diploid organisms, yet many fundamental processes are not yet understood. Among these are the relative importance of known processes, an integration of evolutionary theory and ecology, and the development of a theory for the evolution of microorganisms. These essays will be of great interest to biologists, philosophers, and historians of science.

On Evolution

EDINBURGH
University
Press

ON EVOLUTION

John Maynard Smith

Dean of the School
of Biological Sciences
University of Sussex

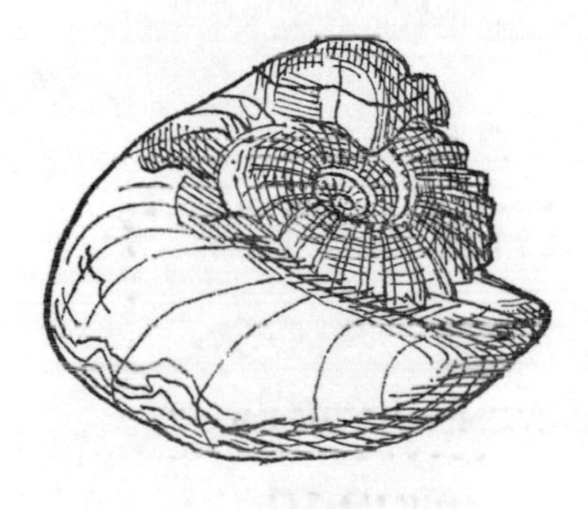

EDINBURGH
University
Press

EDINBURGH UNIVERSITY PRESS
22 George Square, Edinburgh
ISBN 0 85224 229 8
North America
Aldine · Atherton Inc.
529 South Wabash Avenue, Chicago 60605
Library of Congress
Catalog Card Number 72–77 391
Printed in Great Britain
by W & J Mackay Limited, Chatham

Contents

Acknowledgements

This book comprises a selection of papers, most of which first appeared in journals or proceedings. All are concerned with evolution theory.

'The importance of the nervous system in the evolution of animal flight' appeared in *Evolution*, 6 (1952) 127–9.

'Evolution and history' appeared in *Darwinism and the Study of Society* (ed. Michael Banton). London: Tavistock Publications Limited 1961.

'The arrangement of bristles in *Dropsophila*' appeared in *Journal of Embryology and experimental Morphology*, 9 (1961) 661–72.

'Eugenics and Utopia' appeared in *Daedalus, Journal of the American Academy of Arts and Sciences*, 94 (Spring 1965) 487–505, in the issue 'Utopia'.

'The status of neo-Darwinism' appeared in *Towards a Theoretical Biology, 2: Sketches*, pp. 82–9 (ed. C. H. Waddington). Edinburgh University Press 1969.

'Time in the evolutionary process' appeared in *Studium Generale*, 23 (1970) 266–72.

'The causes of polymorphism' appeared in *Symposium of Zoological Society*, 26 (1970) 371–83.

'The origin and maintenance of sex' appeared in *Group Selection*, pp. 163–75 (ed. George C. Williams). Chicago: Aldine Atherton Inc. 1971.

These papers are reproduced by permission of the original publishers, to whom grateful acknowledgement is made.

The essay on 'Game theory and the evolution of fighting' was specially written for this book. I would probably not have had the idea for this essay if I had not seen an unpublished manuscript on the evolution of fighting by Dr George Price, now working in the Galton Laboratory at University College London. Unfortunately, Dr Price is better at having ideas than at publishing them. The best I can do therefore is to acknowledge that if there is anything in the idea, the credit should go to Dr Price and not to me.

JOHN MAYNARD SMITH

Introduction

The publication of this collection of essays on evolution theory provides both an occasion and a stimulus for taking stock. What is the state of evolution theory today? How is it likely to develop? What problems are worth working on? An additional reason for asking these questions is that there appears to be a widespread conviction that there is something rotten in the state of evolution theory, and that profound changes in the theory are imminent. Let me hasten to add that this conviction, although widespread, is confined to those who do not work in the field of population genetics. Population geneticists tend to hold views similar to those held by physicists at the end of the last century: the fundamentals are known, and all that remains is to work out the details. Who is right? On the one hand, one could argue that non-specialists are reluctant to accept that they are the product of a process as mindless and mechanistic as natural selection, and that their ignorance permits them to hope that the concept is false. Certainly the odd enthusiasm for Teilhard de Chardin points to this explanation. But on the other hand, it may be that geneticists are blinded by professional blinkers from seeing the inadequacies of their own subject. There are plenty of other examples in the history of science of pride going before a fall.

I do not know which of these pictures is more nearly correct, although as a professional geneticist I naturally lean towards the former. In the essay on 'The status of neo-Darwinism' I tried to evaluate evolution theory from the viewpoint of a particular philosophy of science, that of Karl Popper. In particular, I was concerned with the problem: to what extent is Darwinism falsifiable? If it is not falsifiable by any conceivable set of observations, then to Popper it would not be a scientific theory at all. In brief, I concluded that there were several kinds of observational evidence which could in principle falsify the neo-Darwinian theory. However, much the most effective way of falsifying the theory would be to falsify the genetic theory on which it is based, and in particular to falsify the Weismannist assumption, popularly but loosely expressed by saying that acquired characters are not inherited.

For reasons I will explain below, a falsification of the Weismannist assumption is today most likely to come from a study of cellular and biochemical heredity—for example, from the study of 'scrapie', a disease which seems to be transmissable in the absence of nucleic acids, or from the study of the immune response, in which individuals 'learn' to make new specific protein antibodies.

If I am right in this, a fundamental rethinking of the neo-Darwinist position depends on a change in our current views on the biochemistry of heredity. At present there seems no likelihood of such a change. However, if it became apparent that current evolutionary theory cannot account for the observed facts of evolution, this would certainly stimulate doubts about the validity of present genetic views. Again, I see no sign of such a breakdown of evolution theory. But at this point I must draw a distinction between two possible meanings of the phrase 'cannot account for'. I can best do this by means of an analogy with another branch of science. Suppose I were to say 'Newtonian mechanics cannot account for the orbit of Mercury', I would imply that it *ought* to be able to account for the orbit, and the fact that it cannot is a good reason for doubting the truth of Newtonian mechanics. But if I said 'Newtonian mechanics cannot account for the internal combustion engine', I would mean merely that if you want to understand internal combustion engines, you will have to develop physical theories additional (but not contradictory) to Newtonian mechanics. Reverting to neo-Darwinism, I do not think there are any observations which require us to doubt the theory, but there are plenty which are not adequately accounted for, and which call for the development of additional theoretical ideas. I will mention one such field of observation at the end of this introduction.

It is interesting to look at the status of evolution theory from the viewpoint of a more recent and more fashionable philosopher of science, T. S. Kuhn. Kuhn's main argument is that science does not grow by continuous addition of new knowledge. Instead, periods of 'normal science', during which scientists work within an accepted framework of methods and assumptions, are interrupted by 'revolutions'. After such a revolution, the practitioners of the new, post-revolutionary science regard different sets of phenomena as being important, interpret old facts in new ways, make different assumptions, ask different questions, and have different standards in deciding what is a satisfactory answer. The difference may be so profound that practitioners of the old and the new orthodoxies cannot understand what the other is saying. Is this an accurate picture of science? If it is, can one detect within the existing orthodoxy of evolution theory the seeds of the coming revolution?

The first thing to be said about Kuhn is that he is saying something about the history of science rather than about its philosophy. To test his ideas, we have to ask whether the history of science is actually like that. Certainly evolution theory at present answers to Kuhn's description of 'normal science'. Although controversies abound, there is agreement as to what constitutes a problem and a solution; there are Kuhnian 'paradigms' illustrating how the subject should be pursued, in such books as Fisher's *Genetical Theory of Natural Selection*, Haldane's *The Causes of Evolution*, or Dobzhansky's *Genetics and the Origin of Species*; there are journals such as *Evolution* and the *American Naturalist*, with accepted standards of refereeing, in which contributions to the subject are published. But are there ever 'revolutions' in the history of genetics or of evolution theory?

There are two main candidates for recognition as revolutions in the history of genetics: the 'Mendelian' revolution, which failed to take place with the publication of Mendel's paper and which had to wait for the rediscovery of Mendel's laws in 1901; and the 'molecular' revolution, initiated in 1952 by the publication in *Nature* of papers by Wilkins and by Watson and Crick outlining the structure of DNA.

The first of these, at least in Britain and in its impact on evolution theory, did have many of the characteristics of a Kuhnian revolution. It is difficult not to conclude that, in the argument between the 'biometrical' school led by Karl Pearson and the 'Mendelian' school of Bateson, neither side really understood what the other was talking about. The Biometricians thought that what mattered in evolution was the accumulation of minute changes; the Mendelians that it was the occurrence of major and discontinuous 'sports'. For the Biometricians the chosen method of research was the statistical analysis of measurements of populations; for the Mendelians it was to follow the segregation of sharp differences in breeding experiments. Most important of all, for Pearson the business of science was to give an adequate mathematical description of phenomena; for the Mendelians it was to reveal the underlying mechanism. In passing, it is worth remarking that few scientists have been so deeply influenced in their work by a consciously worked out philosophy of science as was Pearson, and few have been so seriously misled by it.

Thus the introduction of Mendelian ideas into evolution theory had many of the features of a Kuhnian revolution. It was later shown, using mathematical methods developed from those of Pearson, that both continuous and discontinuous variation could be accounted for by the Mendelian mechanism. But at the time the contradiction between the two sets of ideas seemed to the participants to be absolute. If the Mendelian revolution answers to Kuhn's description, what of the later molecular revolution? I feel a greater confidence in answering this question, because it is a revolution which has occurred during my working life as a biologist.

First, there are ways in which the molecular revolution was indeed a revolution in genetics. As a result, Drosophila was largely replaced as a favoured object of study by bacteria and phage: before the revolution, an explanation was complete if couched in terms of factors or genes of unknown nature and composition, whereas today an explanation is incomplete until the chemical nature and mode of action of the postulated factor is known; on a more mundane level, there has been a struggle for grants and appointments between practitioners of the old and the new. But in more important ways the molecular revolution differs from the Kuhnian picture. The new molecular ideas were not seen by classical geneticists as in any way contradictory to their own ideas; instead they were seen as a natural and long-awaited extension of them. Some of the 'founding fathers' of classical genetics had foreshadowed the molecular revolution; for example, H. J. Muller had speculated about template reproduction of genes, and J. B. S. Haldane had continuously stressed the need for a chemical interpretation of genetics. My own reaction to the double helix was, I suspect, fairly typical of rank-and-file geneticists trained in the pre-molecular era; I did not then (and do not now) follow in detail the methods which led to a solution of the problem. But

I accepted the solution at once, because the phenomenon of complementary base pairing provided a mechanism for gene replication.

My point then is that, so far from being regarded as contradictory to or subversive of classical genetics, the new molecular ideas were accepted by classical geneticists because they provided a chemical explanation for the basic assumptions of Mendelian genetics. Interestingly, the clearing up of anomalies in classical genetics was not a major reason for the acceptance of molecular genetics. Thus Kuhn suggests that it is a sign of the coming revolution within normal science that 'anomalies'—that is, phenomena which cannot be explained within the accepted framework—should accumulate. There were of course anomalies within classical genetics in 1952, but they were not immediately cleared up by molecular genetics. Perhaps the most important anomaly was the phenomenon of 'adaptive enzymes', whereby a bacterial strain starts to produce a specific enzyme only when the substrate for that enzyme is present, a phenomenon which led Hinshelwood to formulate a non-Mendelian and fundamentally Lamarckian theory of heredity. Adaptive enzymes were anomalous because it seemed that, in contradiction to the Weismannist assumption, an adaptation was being inherited. This phenomenon remained unexplained for some time after the general acceptance of the basic concepts of molecular genetics, and was not fully explained until the discovery of 'regulator genes' by Monod and Jacob in 1959.

Within evolution theory, molecular biology has raised new problems concerned with the evolution and variation of proteins, with the evolution of genetic mechanisms, and with the nature of mutation. But it has proved to be an addition to the pre-existing theory rather than a revolution in it. This has been most obviously true for the fundamental Weismannist assumption which underlies neo-Darwinism. In its pre-molecular form, this assumption stated that although genes can influence the development of the body or 'soma', changes in the soma cannot alter genes in an adaptive way. Since 1952, this has been replaced by the 'central dogma' of molecular biology, which states that information can flow from nucleic acids to proteins, but not from proteins to nucleic acids. The correspondence between these two ideas is obvious once it is realized that genes are made of nucleic acid, and that they act by specifying the kinds of proteins a cell can produce. It is for this reason I suggested earlier that a falsification of Weismann's assumption will come, if it comes at all, from biochemical genetics.

Does it follow from this that all that is left for evolutionists to do is to clear up the loose ends? I do not think so, for two reasons. First, some of the problems which remain within the field traditionally covered by evolution theory are by no means trivial ones. Although it seems unlikely that any wholly new evolutionary processes remain to be discovered, at least for the evolution of diploids, the relative importance of different processes is still a matter of debate. The relative importance of selection and of 'genetic drift', due to chance events in finite populations, in producing changes in gene frequency, has been argued since the early work of Fisher and Wright. It has arisen again in a new and intriguing form in the study of protein evolution, and is the subject of one of the essays in this book. A more recent controversy concerns the relative importance of individual and group selection: different aspects of this controversy are discussed in the essays on 'Game theory and the evolution of fighting', and on 'The origin and maintenance of sex'.

I will mention two other fields in which important problems remain to be solved. The first is the consequence for population genetics of the fact that genes are linked on chromosomes. Here the difficulties of mathematical analysis are very great, and have effectively limited progress until recently, when the advent of large computers has encouraged people to return to the problem. Finally, and perhaps more fundamental than any of them, we need to develop a theory of the evolution of micro-organisms. The novelty here is in the 'parasexual' methods of exchange of genetic material, which make it possible for genes to be exchanged between quite distantly related organisms. I suspect that to cope with this we may have to reformulate our models of natural selection. Our present models are concerned with individual animals or plants, and their probabilities of survival and reproduction; the entities which multiply, which vary, and on which selection acts are these individuals. In the evolution of micro-organisms, it may be that the appropriate entities to consider are the genes themselves.

The second and more important reason why evolution theorists have more to do than clear up loose ends is the need to extend the theory or to develop new ones to cope with new aspects of biology. The field in which this is happening most actively at present is in the field of ecology. Ecology has concepts proper to itself, which are not derived from evolution theory: examples are the concepts of density-dependent population regulation, of ecological succession, and of competitive exclusion. All the same, evolution theory has something to contribute. It can supply 'restraints' which any ecological theory must satisfy, in much the same way that Newtonian mechanics, although itself unable to account for the internal combustion engine, can insist that any successful account should satisfy the laws of conservation of mass and momentum. The essential restraint applied by evolution theory is that species should consist of individuals whose properties can be understood as arising by natural selection.

We are becoming uncomfortably aware that man has disturbed the balance of nature, and that this disturbance is so great that it is doubtful whether we can find a new point of equilibrium before disaster overtakes us. The trouble with this awareness is that we have very little idea of what maintained the balance of nature in the first place, or even of the extent to which such a balance existed. The crisis of the environment has of course a political dimension as well as an ecological and an evolutionary one. Nevertheless there is urgent need for a theory of population biology which will include both the ecological and the evolutionary time scales. There are promising signs that such a theory is being developed.

Game Theory and the Evolution of Fighting

The Problem

As everyone knows, male deer have branched antlers. During the breeding season, two stags fight by lowering their heads so that their antlers interlock. Each then attempts to force the other backwards, until at last the weaker is forced to break away and flee. Because of the branching structure of the antlers it is rare for a stag to be pierced by its opponent's antlers. Occasionally, however, a stag grows antlers without branches; such a stag may wound and kill its adversary. Now victory in these fights ensures to the victor the possession of a harem of hinds, and so increases his expected contribution to the next generation. Why should natural selection have favoured a device—the branching of antlers—which appears to reduce the chances a stag may have of winning fights?

This is not an isolated problem. Males of a species often indulge in such 'conventional' methods of fighting, even though by breaking the conventions they would increase their chances of victory. Bighorn rams leap at one another so that they meet head on. It would seem far more effective to wait until an opponent's back was turned and charge him in the flank. In some species of fish, rivals fight by seizing one another by the jaws and either pulling or pushing; in these species the mouth is covered by leathery skin, so that a bite anywhere else would be more likely to cause injury. In the fence lizard, a male seizes his opponent by the head, which is heavily armoured; after a short period of wrestling, he lets go and allows his opponent to have a turn. Many alternate bouts of this kind may take place before one gives way.[1]

Even more commonly, the result of a conflict[2] is decided without physical contact, by visual or other display. In such cases, the 'loser' accepts defeat without even putting the matter to a physical test of strength.

Before considering the orthodox explanation for these facts, it is worth remembering that conventional fighting is found in man as well as among animals. Few combats are more conventional than a boxing match, with its padded gloves, rules against hitting below the belt, rest periods between rounds, and so on. It would be a mistake to suppose that such conventional fighting is confined to civilized societies; the natives of Tierra del Fuego, whom Darwin regarded as the most primitive of mankind, indulged in a highly formalized type of wrestling.[3] Nor is it true that conventional fighting in man is confined to situations in which nothing much depends on the outcome, or in which there is a third party to enforce the rules. In mediaeval times, disputes were commonly settled by a fight between champions. This was true for disputes between communities as well as between individuals; for example, in 1085 single combat decided whether the Toledo liturgy should be read in Latin or Mozarabic.

In these human examples, it is reasonable to suppose that the acceptance of conventional restraints depends on the rational calculation that it is better to minimize the risks one runs in the event of defeat. In the case of animal conflicts, an explanation in terms of rational calculation is less acceptable. The explanations most commonly accepted for the ritual and conventional nature of animal conflict are twofold. First, an individual which attacks other members of its species without preliminary signals of intent is unlikely to obtain a mate, and if it does may well kill its children. This is a satisfactory explanation of the existence of ritualized courtship and appeasement displays, but is less satisfactory as an explanation of the ritualized or conventional nature of the conflicts between rivals for mates, food, territory, and so on. Second, in a species in which there are no conventional restraints on the methods used in fights, many individuals will be injured, and this will militate against the survival of the species.[4]

The second of these arguments raises some difficulties. Thus consider a conflict between two individuals, A and B. Suppose that A, by ignoring conventional restraints, injures B and forces him to give way. This will increase A's probability of leaving offspring (supposing the conflict to be about something of value to survival), and decrease B's chance. Thus at the individual level, selection will favour A, who ignores conventional restraints, at the expense of B, who obeys them. It is true that if A's type of behaviour is common, it may be bad for the survival of the species or population as a whole. Thus there is a conflict between 'group selection' favouring conventional behaviour and 'individual selection' against it. It follows that if we argue that conventional methods have evolved because they reduce the risk that individuals will be injured, we may tacitly be assuming that group selection is more effective than individual selection. For reasons given in the next section, this is probably not so. The main purpose of this paper is to see whether an explanation of conventional fighting can be given in terms of selection at the individual level. But first, I want to explain in a little more detail my reluctance to accept group selection as an explanation.[5]

Group selection and kin selection

Most discussions of natural selection, and almost all mathematical models of it, are concerned with individual selection. A population consists of individuals of two (or more) different genotypes, say *A* and *a*. On the average, *A* individuals leave more offspring than *a* individuals. Depending on the particular mechanisms of inheritance, this will cause a calculable change in gene frequency each generation. This is 'individual selection'. In contrast, suppose that a species were divided into a number of groups, reproductively isolated from one another, some consisting entirely of *A* individuals and some of *a* individuals, and suppose also that from time to time populations become extinct and that other populations split to form two. It might be that *A* groups are more likely to survive and split than *a* groups. If so, in time all groups will consist entirely of *A* individuals.

This remains true even if, in a mixed group of *A* and *a* individuals, *a* replaces *A* by individual selection. To give an example, if *a* individuals are cannibalistic, it might be that *a* groups are liable to extinction, but that in mixed groups *a* would win over *A*.

The process whereby all groups become *A* groups is called group selection. Note that it will not work if migration from group to group is at all common, because groups of *A* individuals are constantly being 'infected' by migrant *a* individuals, and any 'infected' group will become an *a* group by individual selection. However, no such migration occurs between different species, so group selection can certainly act at the species level, by the extinction of some species and the survival of others. It may also act within a species which (because of the spatial distribution of available habitats) is divided into isolated groups for much of the time.

Although group selection is a possibility, it is rather ineffective compared to individual selection. This is most easily seen by considering the fate of a 'harmful' mutation. If a mutant is individually harmful, each new mutation can be eliminated from the population by a single selective death. If a mutant is individually beneficial but harmful to the group, each new mutation will be spread through the population (or species, if there is no division into reproductively isolated populations) and can be eliminated only be the extinction of the whole population or species.

For this reason, I am reluctant to accept group selection as an explanation for an adaptation if there is a possibility of explaining it in terms of individual selection. In the present context, this means that an explanation of a particular behaviour pattern of fighting which amounts to saying that it is good for the species but bad for the individual should be accepted only as a last resort.[6]

It is also relevant to the evolution of fighting to consider 'kin selection', which depends on the fact that individuals have many genes in common with their relatives. Thus a behaviour pattern which reduces the probability of survival of an individual, but at the same time increases (to a greater degree) the probability of survival of close relatives, may increase in frequency. The classic example of kin selection is in the social insects, in which some individuals (workers) refrain from breeding and by doing so increase the chance of survival of their fertile siblings.

In the context of the evolution of fighting, kin selection is relevant in two ways. First, within a family or closely related group, kin selection will reduce the severity of conflicts. To natural selection, to kill a twin, or to kill two brothers, is as bad as to kill oneself. To give an extreme example, in the native hen, *Tribonyx mortieri*, it is common for two brothers to share a wife,[7] although if one were to drive out the other he would father all the children born to the female instead of only half. This makes sense only if one remembers that the losing brother would have transmitted many of the same genes as the winner, so that by driving him out the winner reduces the number of his own genes transmitted to the next generation.

The other relevance of kin selection is to the evolution of

warfare. *'Dulce et decorum est pro patria mori'*, but only if one's *patria* consists of individuals with many of the same genes as oneself. Outside our own species, warfare is most clearly developed among the social insects. I was surprised to discover that it is also found among spotted hyaenas.[8] A tribe of hyaenas will occupy a well-defined territory, and attack members of other tribes who venture close to the boundary. An individual hyaena close to the tribal boundary may be attacked by members of the neighbouring tribe, and may be killed unless rescued by the arrival of reinforcements. It is rare (but apparently not quite impossible) for an individual born in one tribe to transfer to another, so that the genetic relationship between the individuals of a tribe is close enough to make warfare between tribes comprehensible.

Selection which depends on how others behave

I remarked earlier that most models of natural selection are of individual selection. A second assumption which is often made is that the relative fitness of two genotypes *A* and *a* does not depend on the composition of the population. This is equivalent to assuming that the optimal phenotype is independent of the phenotypes of the rest of the population. It is questionable whether this is ever absolutely true, but it is often near enough for practical purposes. In a desert animal, any improvement in the ability to conserve water is likely to be favoured by selection, regardless of the composition of the population. But in most cases survival depends on competition or collaboration with other members of the species, and therefore the optimal phenotype depends on the rest of the species. This is usually the case for behavioural characteristics.

When a behavioural pattern involves communication between different individuals, it is obvious that the optimal behaviour depends on how others behave. This is particularly relevant in the evolution of courtship, when there must be a joint evolution and adaptation of the signals given by one partner and the response of the other. However, courtship between two conspecific individuals is not an example of conflict, since both partners have a common interest in mating. It is not therefore relevant to the subject of this essay, which is the evolution of behaviour appropriate to situations in which there is a conflict of interests.

The typical conflict situation is one in which two members

of a species compete for some resource—a mate, a territory, food, a nesting site or nesting material, and so on. There is however one other conflict situation worth discussing; this is in the 'choice' of a sex ratio among one's offspring.[9] Thus imagine that the males in a sexually reproducing species each produce the same number of offspring and similarly for the females, but that they can influence the sex of their offspring. Each member wants to maximize the number of its descendants, relative to the number of descendants produced by its competitors. (Less anthropomorphically, we suppose there are genes which influence the sex ratio among the offspring of an individual carrying them, and ask which gene will be transmitted to most descendants.) What sex ratio should an individual produce? The answer clearly depends on what sex ratio others are producing. If the rest of the population are producing males, then an individual producing only females will maximize the number of his (or her) grandchildren, and *vice versa*.

This is an example of a conflict, because an individual can only increase his relative contribution to future generations by reducing the contribution made by others, and because his optimal 'strategy' (that is, to produce sons or daughters or some appropriate mixture) will depend on what strategy others are adopting.

Game Theory

I decided to learn something about game theory because I had become interested in the evolution of threats and aggressive behaviour, and for quite different reasons in the evolution of the sex ratio. I started out knowing only that game theory was concerned with conflicts and that there were said to be some powerful theorems available. It is tempting to learn a new branch of mathematics, in the hope that it will contain ready-made solutions to one's problems. In practice there always turns out to be some good reason why the theorems cannot be applied, as it does in this case. But, and this too has proved to be true of game theory, some of the concepts and techniques can be made use of.[10]

I cannot in a few pages present even the elements of game theory, but I hope to say enough to indicate what kind of theory it is. Game theorists are interested in conflicts between two or more 'players'. It is essential that there be a conflict of interest, so that what is good for one player is, at least sometimes, bad for the other. Before the theory can be applied, one must be able to make a complete list of the 'strategies' available to each player. By a 'strategy' is meant a specification of what a player will do. Thus a 'strategy' for the first player at noughts and crosses might start out. 'I will put an o in the centre square. If he puts his first x at the centre of a side, I will put my second o in a corner next to it', and would have to go on to say what I would do in response to all my opponent's possible second and third moves. Obviously, a listing of all possible strategies for both players at even such a trivial game as noughts and crosses would be laborious, although possible; for chess it would be impossible, although a finite but immense list of strategies exists. For the 'sex ratio game' a list of strategies would simply be a list of possible sex ratios—for example 1:0, 0·9:0·1, 0·8:0·2, . . ., 0:1.

In addition to a complete list of strategies it must be possible to calculate the outcome of the game for every possible pair of strategies, and to ascribe a 'utility' to each player for each outcome. To illustrate what this means, for noughts and crosses it is easy to calculate, for any specified pair of strategies (one for each player), who will win. Each player might ascribe +1, 0, and −1 respectively to a victory, a draw, or a loss for himself. Unfortunately there is nothing either sacred or obvious about these utilities. One player might feel there was nothing particularly creditable about winning, since it was possible only because his opponent was careless, but that it was deeply shameful to lose; if so he might ascribe utilities $+\frac{1}{2}$, 0, −10 to the three outcomes.

The ascription of 'utilities' to outcomes is a major difficulty in applying game theory to actual problems—fundamentally because it requires that one puts incommensurables (the shame of losing, the financial gain of winning) in a single numerical scale. Fortunately, however, the difficulty is minimal when applying game theory to evolutionary problems.

To fix ideas further, suppose two players, John and Eric, play the game 'rock–paper–scissors'. The rules are as follows. The two players extend their hand simultaneously, either clenched to represent a rock, flat to represent paper, or with the first two fingers extended to represent scissors. If the two players make the same symbol, no money changes hands. If different symbols, then rock beats scissors, scissors beat paper

paper beats rock, and the loser pays one penny to the winner. We can list the possible strategies, and the utilities (equated to cash) in a 3×3 'pay-off matrix' as shown in table 1.

Table 1

		ERIC		
		rock	scissors	paper
JOHN	rock	0	+1	−1
	scissors	−1	0	+1
	paper	+1	−1	0

The pay-offs are to John: for Eric all the signs must be changed.

To the game theorist, this is a complete representation of the game. He then makes two assumptions, and on the basis of those assumptions can draw two conclusions. First the assumptions:

(1) John should play so as to maximize his gains; *but*
(2) he must bear in mind that Eric will play so as to maximize *his* gains.

John should therefore adopt the so-called 'maximin' strategy, which maximizes his gains, assuming Eric is doing all he can to minimize them.

These are the assumptions. The game theorist will then attempt to answer two questions:

(1) What is John's best strategy?
(2) If he adopts this strategy, what is the least he can expect to win, assuming Eric does his best to minimize John's winnings?

All this sounds (and is) fairly trivial. It has been presented mainly in order to introduce the concept of a pay-off matrix, because this is the tool I want to use later. But before leaving game theory proper, there are a few other points to be made.

The strategies 'rock', 'scissors', and 'paper' are called 'pure strategies'; there are also 'mixed strategies'. A typical mixed strategy would be 'with probability $\frac{1}{4}$, rock; with probability $\frac{3}{4}$, scissors'. Only in rather special cases will John's best strategy be a pure strategy; usually he should adopt a mixed strategy. (In case you are curious, in this case he should adopt a mixed strategy of $\frac{1}{3}$ rock, $\frac{1}{3}$ paper, $\frac{1}{3}$ scissors, and he can expect to break even. But game theory does come up with more interesting answers than this.)

The rock–scissors–paper game, as outlined, is a 'zero-sum' game. What John wins, Eric loses; Eric's pay-off matrix is the same as John's with the signs changed. Not all games are 'zero-sum'. Suppose John and Eric played the rock–scissors–paper game with the agreement that the loser shall be obliged to commit suicide, having first given £10 to the winner to compensate him for his bereavement. This is no longer a zero-sum game. Unfortunately, most games are non-zero-sum, but most of the successes of game theory have been in the analysis of zero-sum games.

The difficulty in the analysis of non-zero-sum games lies in the possibility of collusion. In a zero-sum game, since what is lost by one party is won by the other, neither side has anything to gain by collusion. But in the non-zero-sum game, John and Eric would be well advised to collude. If they agree beforehand that each will display the same symbol—say a rock—then neither runs the risk of suicide. Even this is not the end of the problem. Suppose they agree each to display a rock. Then, if the game is to be played once only, John may be tempted to earn himself £10 by displaying paper.

Happily, in the analysis of evolutionary 'games', an additional assumption can be made which enables us to handle non-zero-sum games. This assumption will be explained later.

Previous applications of game theory to evolution

Before applying the ideas of the last section to the evolution of fighting, it will be interesting to review some previous applications of game theory to evolution. Kalmus[11] discussed in general terms the ways in which game theory might be applied to animal behaviour, but did not attempt to solve any particular problems. Lewontin[12] wrote a paper entitled 'Evolution and the theory of games', in which he was concerned with the question: what genetic mechanisms will maximize the chances of survival of a species? In his analysis, one of the two 'players' is the species, and the other is nature. The strategies open to the species are the various genetic mechanisms it might adopt—for example, sexual *v.* asexual reproduction, genetic polymorphism *v.* individual physiological adaptation. The strategies open to nature are the various environments the species may have to meet.

The idea of treating nature as one of the players in a game did not originate with Lewontin. Some of the examples of games in elementary texts on game theory make this assumption. In general, the assumption is quite unwarranted. It is a basic assumption of game theory that one's opponent will behave so as to maximize his own gains—and therefore, in a zero-sum game, so as to minimize yours. To apply this assumption when one's opponent is 'nature' is to assume the truth of 'Sod's law', which states that if a thing can go wrong it will—a law with the familiar lemma that if one drops a piece of bread and marmalade it will fall marmalade side down. If one cannot make this assumption, then game theory does not apply.

The attempt to apply a mathematical theory to nature by introducing Sod's law is reminiscent of the attempt to apply probability theory to hypothesis-forming by introducing Bayes' postulate. Bayes' postulate assumes that if you do not know which of several possible states of affairs is the cause, then they are all equally likely; Sod's law assumes that the state which is the worst for you is the case. There is no general reason to accept either postulate.

Lewontin is aware of this difficulty, and goes a long way to meet it in the case of evolutionary games against nature. He argues that 'success' for the population requires long-term survival, and this requires survival not only in good years but also in bad ones. Hence 'the population's optimal strategy is to keep the local probability of survival as high as possible under the worst combination of states of nature'. ('Local' in this quotation means local in time—for example, in a single generation.) In other words, although Sod's law is not always true, a species must behave as if it were; a species should adopt a maximin strategy. With this assumption, Lewontin is able to make some headway with his problem.

The other application is more straightforward, and is due to Hamilton.[13] It concerns the evolution of the sex ratio. In some groups (for example, Hymenoptera and some mites), fertilized eggs develop into females and unfertilized eggs into haploid males. The females store sperm, and so can 'choose' the sex of their offspring by laying fertilized or unfertilized eggs. Many such animals are free-living as adults but parasitic as larvae; the adult female searches for a host (often a caterpillar) and lays eggs in it. These eggs develop at the expense of their host. When they emerge as adults, it is common for mating to take place at once, after which the males die and the females disperse to look for new hosts. When a female finds a caterpillar, what ratio of female to male eggs should she lay in it? If there are no other females around, the answer is obvious. Since one male can mate with many females, she should fertilize all her eggs except one. In this way she will maximize the number of her grandchildren.

But what if there are other females around? Then we are faced with a problem in game theory, in which each female must select a sex ratio which will maximize her contribution to future generations. Her choice depends on what others are doing. For example, if all other females are laying, say, 20 female and 1 male egg per caterpillar, it might pay a female to lay only a few eggs in each caterpillar, all of them male. The correct strategy depends on many things, and in particular on whether the fertility of a female is limited by the number of eggs she can lay or by the number of caterpillars she can find to lay them in. But Hamilton is able to show that the sex ratios actually adopted can plausibly be thought of as optimal from a game-theory standpoint.

Game theory and fighting—a simple model

The time has now come to apply game theory to the problem of intra-specific fighting. The model which I will present is highly schematic. It leaves many things out of account (for example, changes of motivation with position in space in territorial animals, previous experience of conflicts with the same opponent, changing behaviour with age, the possible risks of attacking a sexual partner). But I think it leaves enough in to be illuminating.

The first step is to produce a pay-off matrix for a conflict between two rivals. It is supposed that each contestant has two 'levels' of fighting available to him:

C, or 'conventional fighting', and

E, or 'escalated fighting'.

C may involve physical contact, or only display. It does not involve actions likely to injure the opponent seriously. In contrast, *E* necessarily involves physical contact, and may lead to injury to the opponent.

Although a contestant at any instant must adopt one or other of these behaviours (or may retreat), he has more than two pure strategies available to him, because he can vary his behaviour according to the behaviour of his rival. There seem to be five pure strategies worth considering:

(1) $C \rightarrow C$ Fight conventionally. Retreat if one's opponent proves to be stronger (or to display with greater vigour) or if one's opponent escalates.

(2) $E \rightarrow E$ Fight at escalated level. Retreat only if injured.

(3) $C \rightarrow E/E$ Start conventionally. Escalate only if one's opponent escalates. Retreat if injured.

(4) $C \rightarrow E/C$ Start conventionally. Escalate only if opponent continues to fight conventionally.

(5) $E \rightarrow r/E$ Fight at escalated level. Retreat before getting hurt if one's opponent does likewise.

This seems to exhaust the single strategies which might prove optimal. The next step is to draw up a pay-off matrix. Before going into details, there is a general point to be made about 'utilities' in evolutionary games. It was pointed out earlier that the ascription of utilities to outcomes is conceptually the most unsatisfactory part of game theory. In the evolutionary context, no such difficulty arises if the game is between two members of a species. The utility of an outcome is simply the contribution that outcome makes to the 'fitness' of the individual—that is, to the expected number of future offspring born to that individual.

In filling in the pay-off matrix, I have made the following assumptions:

(1) The contest is over some resource (for example, food) which will contribute to the chance of survival or fertility of the winner. This resource is valued at $+2$.

(2) Two contestants adopting the same behaviour have an equal chance of winning, and so have an expected gain of $+1$.

(3) Two contestants who both indulge in escalated fighting run a risk of serious injury. This risk is valued at -11; the exact value is unimportant to the argument, but it is important that it is absolutely greater than the expected gain from winning.

With these three assumptions, the pay-off matrix in table 2 is obtained. In each case the figure below and to the left is the utility of the outcome to contestant A, and the other figure to contestant B. To illustrate how the matrix is derived, consider two examples:

(1) A adopts $E \rightarrow E$; B adopts $C \rightarrow E/E$. Both escalate. Hence

both have a utility of +1 (for an even chance of winning) −11 (for the risk of injury) = −10.
(2) A adopts $C \to E/E$; B adopts $E \to r/E$. B starts to escalate, whereupon A retaliates, and B runs away. A gets the resource, worth +2, and B gets nothing.

Table 2

	$C \to C$	$E \to E$	$C \to E/E$	$C \to E/C$	$E \to r/E$
$C \to C$	+1 +1	+2 0	+1 +1	+2 0	+2 0
$E \to E$	0 +2	−10 −10	−10 −10	0 +2	0 +2
$C \to E/E$	+1 +1	−10 −10	+1 +1	−10 −10	0 +2
$C \to E/C$	0 +2	+2 0	−10 −10	−10 −10	+2 0
$E \to r/E$	0 +2	+2 0	+2 0	0 +2	+1 +1

Given this pay-off matrix, how should A behave? At first sight, it might seem that he should adopt either strategy $C \to C$ or $E \to r/E$, since only in these two cases can he avoid the risk of being injured, and, assuming that his opponent makes a similarly conservative choice, A should prefer $E \to r/E$, which does better against both alternatives. But this does not solve the problem, because, if B can 'calculate' that A will adopt a strategy $E \to r/E$, it will pay him to adopt the otherwise dangerous strategy $E \to E$ or $C \to E/E$. In evolutionary terms, if a population consists of individuals which adopt strategy $E \to r/E$, any mutant individual adopting $E \to E$ will be favoured by selection. We are in trouble, because the game is a non-zero-sum game.

However, we can solve the problem, because we are not seeking for A's optimal strategy in game theory terms. We are seeking for an 'evolutionarily stable strategy', or ESS for short. A strategy qualifies as an ESS if, in a population in which most individuals adopt it, there is no alternative strategy which will pay better. Thus $E \rightarrow r/E$ is not an ESS, because in a population of $E \rightarrow r/E$ individuals, strategy $E \rightarrow E$ would pay better; one might as well fight to the finish if one's opponent is sure to run away. But neither is $E \rightarrow E$ an ESS, because in a predominantly $E \rightarrow E$ population, $E \rightarrow r/E$ would pay better; if everyone else fights to the death it is wiser to run away.

An inspection of table 2 shows that only one of the five pure strategies, $C \rightarrow E/E$, is an ESS. In a population whose members fight conventionally, but which escalate if their opponent escalates, there is no alternative strategy which would pay better. There is however a strategy, $C \rightarrow C$, which would do equally well.

It follows that if a population adopting strategy $C \rightarrow E/E$ were to evolve, the behaviour would persist and departures from it would be eliminated by selection, except that $C \rightarrow C$ individuals might increase in frequency by mutation. This increase in $C \rightarrow C$ individuals would be checked if the typical members of the population occasionally escalated fighting without provocation. This behaviour might not pay if encounters were unique, but would pay off if the same pairs meet repeatedly, since in this case $C \rightarrow C$ individuals would in time be detected.

Our model therefore leads to the conclusion that the evolutionarily stable state of the population is one in which individuals usually fight conventionally, but escalate if their opponent escalates, and occasionally escalate without provocation.[14] In other words, individual selection can account for the evolution of behaviour patterns which minimize injury, so that there is no need to invoke selection between groups as an explanation.

Ritualized conflicts

The model considered in the last section leaves many things out which would need to be considered in a full analysis of the evolution of threat and fighting. However, all good models in science leave out a lot. A model which included everything would be too complicated to analyze. The purpose of the model was not to give a complete picture, but to answer a particular problem, that is, why do animals refrain from injuring their opponents? The model has served this purpose fairly well. Unfortunately, there is a difficulty which I have avoided so far, but which must be faced if one is to have any confidence in the model, even for its own restricted purpose.

The difficulty lies in the difference between what I have called conventional fighting and purely ritual conflicts, conducted by visual displays, singing, and so on. It is an essential feature of the model that two individuals which refrain from escalated fighting can nevertheless settle a conflict, so that one of them wins and the other loses. In a conventional fight, this is reasonable; whichever is the larger and stronger can win, without injuring the loser. But in a ritualized conflict, why should either side yield? Strictly, why should it be selectively advantageous to yield in a ritualized conflict? One possible answer is that there is always a risk that an opponent who does not get his own way by display may resort to physical attack. But more often the disadvantage of continuing to display is simply that it wastes time and energy. A bird which spent too much time defending or extending its territorial boundary would have too little time for nest-building, feeding, and so on.

In ritual conflicts therefore both participants stand to gain if they can win the conflict, but they also stand to gain if they can settle it and to lose if it continues. This bears an analogy to human 'bargaining' situations in which both sides stand to lose if they cannot reach agreement. It is helpful to consider a human example in some detail. Suppose that an employer and a trade union are negotiating a wage claim. The union has demanded a 15% rise, and the employer has offered 5%. Suppose also that the employer would actually be willing to pay 12% rather than have a strike, and that the union would settle for 8% rather than have a strike. Needless to say, neither side will admit openly to these figures. If the union admitted it would settle for 8% then it would be unlikely to get more, and the same applies in reverse to the employer.

A good deal has been written about the theory of such conflicts,[15] but much of it is irrelevant to animal conflicts because it relies too heavily on the assumption of rationality. But one principle is common to human bargaining and animal conflict. It is necessary to give the appearance of being willing to hold out for the maximum, while in practice being willing to settle for less. An individual who reveals his willingness to compromise too easily will be forced to compromise at a point unfavourable to himself, whereas an individual who is unwilling to compromise at all runs the risk of failing to settle the conflict, when a settlement would pay him.

How does this apply to animal conflicts? In most conflicts an animal will be under competing motivations to continue or to retreat. The strengths of these movitations will depend in part on immediate circumstances, in part on previous experience of his opponent, and in part on genetic factors, which will in turn depend on the past selective advantages of the two possible courses of action. However, if the earlier analysis is correct, it will not pay an animal to reveal the exact state of its motivation, any more than it pays a negotiator to reveal at what level he will settle. Instead, natural selection will favour a sharp switch in behaviour at some threshold level of motivation. So long as there is a motivational balance in favour of continuing a conflict, display should be continued at full intensity; once the balance shifts in favour of retreat there is nothing to be gained by beating about the bush.

This analysis, like the previous one, is greatly oversimplified. Two points do emerge, however. First, purely ritual conflicts can be settled, since it is selectively disadvantageous to continue them too long, because of time-wasting and perhaps because of the increasing risk of escalation. Second, the logic of such conflicts leads one to expect what Waddington[16] would call a canalization of behaviour, with a sharp switch from aggressive to submissive behaviour at some threshold of motivation. The latter conclusion is particularly interesting. After reaching it by the preceding line of argument, based on game theory, I was delighted to find that a principle of 'typical intensity' has been formulated by Morris,[17] according to which ritualized acts are performed, if at all, at a fixed level of intensity regardless of the degree of motivation. Morris explains this as a method of avoiding ambiguity. My own explanation, in the case of conflict situations, is somewhat different, since it implies that the advantage of fixed intensity of display is that it conceals the exact level of motivation. In a 'bargaining' situation, ambivalent feelings should be concealed behind an unambiguous front.

To what extent do animals involved in ritualized conflicts

display with typical intensity, as predicted by game theory, and to what extent do they vary the intensity of their display as an indicator of their state of motivation? Cullen[18] quotes the case of the sword tail, *Xiphophorus helleri*, in which the male has an S-posture threat display, with its tail bent slightly towards its rival and its head slightly away. Two rivals will maintain this typical form, until one attacks or the other flees. This is precisely what game theory would predict. Unfortunately, Cullen states that other species of *Xiphophorus* have not evolved a 'typical form' to the same extent. Perhaps the most carefully analyzed case is that of ritualized conflicts in the Siamese fighting fish, *Betta splendens*. Ritual conflicts between males are usually followed by escalated fights, in which one or both rivals may be seriously injured. Conflicts between females however often end (typically after 5–15 minutes) with the surrender of one fish, without escalated fighting. Simpson[19] followed such conflicts in detail, measuring the frequency and timing of particular components of the ritual. He found no significant difference between the frequencies with which eventual winners and eventual losers performed particular acts, except during the last 2 minutes of a contest, when the eventual winner could be recognized from the fact that her gill covers were erected for a larger proportion of the time. The fact that the winner could not be distinguished from the loser until close to the end of a contest fits well with the prediction from game theory.

One other idea from the study of human bargaining is worth discussing. This is the attraction of a 'unique point' on a bargaining scale. Thus in the example of the trade union and employer given above, it is quite likely that the eventual settlement will be at 10%, which is unique both in being a round number, and in being half way between the initial positions. What is certain is that if at some stage in the negotiation the union announced that it would settle for 10·1%, then it would settle for 10%; oddly enough, if the union announced it would settle for 10%, it could convincingly refuse to retreat further.

The logic behind this apparently illogical situation appears to be as follows. Both partners to the negotiation want to settle, and there is a range of outcomes which both would accept. In the absence of any relevant criterion which they can use to decide at which position on the scale they should settle, they are willing to accept some point merely because it is unique. Clearly, this concept can be applied only when the resource which is being competed for is divisible. An obvious application is to territorial disputes between animals; it seemed worth enquiring whether there is an animal equivalent of the 49th parallel. In fact, it appears that there is. Animals will often settle on some feature which has no relevance to survival as a territorial boundary; for example, mice will adopt a chalk mark on the floor of their cage as a boundary.

I do not want to make too much of this aspect of ritualized displays; mice can settle on a territorial boundary without the help of a chalk mark. Two points are important for my argument. First, ritualized conflicts can be settled, because not to settle wastes time and energy and may risk injury. Second, game theory predicts a 'canalization' or 'typical intensity' of ritualized acts used as threats; there is evidence for such canalization in some cases, but it is not universal, and the reasons for the departure from the principle of canalization are not clear.

Conclusion

In this essay I have drawn analogies between human and animal conflicts, and applied to the evolution of animal behaviour concepts of game theory which were developed to analyze human conflicts. In general, I am somewhat distrustful of analogies of this kind, so in conclusion I want to say something in defence of my use of them. When an analogy is drawn between a human and an animal action—for example, between a boxing match and a fight between two stags—one of two very different points may be being made. First, it may be being suggested that the physiological mechanisms underlying the two actions may be similar; for example, that the same hormones may be important in the two cases. This is the type of conclusion from animal-human analogies that I distrust. There may or may not be physiological similarities between human and animal aggression, but nothing I have said in this essay is intended as evidence for such similarity.

The second reason for using an analogy is the belief that there is a *logical* similarity between two processes. Most uses of analogy in science are of this kind.[20] For example, engineers used to draw an analogy between the stresses in a beam and the shape of a soap bubble, because the equations describing the two situations are identical, and not because they thought that beams are made of soap. The analogies in this essay are similar in kind. I think there is often a logical similarity between the role of human reason in optimizing the outcome of a conflict between men, and the role of natural selection in optimizing the outcome of a fight between two animals.

Notes and References

1 The last two examples are from K. Lorenz (1966) *On Aggression*. London: Methuen.

2 Here, and throughout this essay, I shall use the term 'conflict' to refer to situations in which there is a conflict of interests between two or more individuals or groups. In the ethological literature, the word conflict most frequently refers to a conflict of 'drives' or 'motivations' within a single individual.

3 E. L. Bridges (1951) *The Uttermost Part of the Earth*. London: Hodder & Stoughton.

4 J. S. Huxley, who was one of the first to recognize the ritualized nature of many animal actions, gave avoidance of injury as the main function of the ritualization of aggression (1966, *Phil. Trans. Roy. Soc. B*, 251, 249–71).

5 The basic idea of kin selection was familiar to the pioneers of evolutionary genetics, Fisher, Haldane, and Wright. Its consequences have been worked out in more detail by W. D. Hamilton (1964, *J. theoret. Biol.*, 7, 17–52). I believe I was the first to use the term 'kin selection' in a paper (1964, *Nature*, 201, 1145–7) pointing out the distinction between kin selection, which requires only that relatives live near one another, and group selection, which requires the species to be divided into reproductively isolated groups.

6 K. Lorenz (1966, *ibid*) is clear about the distinction between group and individual selection, and aware that the latter can produce results which are bad for the species. He seems less aware of the relative ineffectiveness of group as compared to individual selection, and therefore accepts without qualms the conclusion that the function of ritualization is to prevent injury. More recently, ethologists have tended to

avoid discussions of function (that is, selective advantage) in favour of a more detailed analysis of physiological mechanisms.

7 J. Maynard Smith & M. Ridpath (1972) *Wife-sharing in the native hen* Tribonyx mortieri. *Amer. Natur.* (in press).

8 H. & V. van Lawick-Goodall (1970) *Innocent Killers.* London: Collett.

9 The consequences of selection on the sex ratio are not easy to see. R. A. Fisher, in *The Genetical Theory of Natural Selection* (1930, Oxford University Press), got close to the correct answer. Extensions to his analysis have been made by R. H. MacArthur (1965, in T. Waterman and H. Morowitz, *Theoretical and Mathematical Biology*. New York: Blaisdell) and by J. M. Campos Rosado and A. Robertson (1966, *J. theoret. Biol.*, 13, 324–9).

10 A good introduction to game theory is given by A. Rapoport (1966, 1970) *Two-Person Game Theory,* and *N-Person Game Theory*. University of Michigan Press.

11 H. Kalmus (1969) Animal behaviour and theories of games and language. *Anim. Behav.*, 17, 607–17.

12 R.C.Lewontin (1961) Evolution and the theory of games. *J. theoret. Biol.*, 1, 382–403.

13 W. D. Hamilton (1967) Extraordinary sex ratios. *Science*, 156, 477–88.

14 The main reason for thinking that this model has some correspondence with the truth is that it is a quite general finding that animals tend to respond to being hurt by escalated fighting. *See*, for example, R. Ulrich (1966) Pain as a cause of aggression. *Am. Zool.*, 6, 643–62.

15 For example, T.C.Schelling (1960) *The Strategy of Conflict*. Cambridge, Mass: Harvard University Press.

16 C. H. Waddington (1957) *The Strategy of the Genes.* London: Allen & Unwin. And earlier.

17 D. Morris (1957) 'Typical Intensity' and its relation to the problem of ritualization. *Behaviour*, 11, 1–12.

18 J.M.Cullen (1966) Reduction of ambiguity through ritualization. *Phil. Trans. roy. Soc. B*, 251, 363–74.

19 M. J. A. Simpson (1968) The display of the Siamese

fighting fish, *Betta splendens*. *Animal Behaviour Monographs*, **1**, 1–73.

20 The nature of analogies in science is discussed in greater detail in the essay on 'Evolution and history' in this book.

The Importance of the Nervous System in the Evolution of Animal Flight

In order to be able to fly, an animal must not only be able to support its weight but must also be able to control its movements in the air. Since animals do not have to learn to fly, or at most need only to perfect by practice an ability already present, it follows that there has been evolution of the sensory and nervous systems to ensure the correct responses in flight. Although no direct evidence on this point can be obtained from fossils, something can be deduced from the gross morphology of primitive flying animals. This can best be done by comparison with the control of aeroplanes, since the latter problem is well understood.

The stability of primitive flying animals

If an aeroplane is to be controlled by a pilot, it must be stable. An aeroplane, or a gliding animal, is stable if, when it is disturbed from its course, the forces acting on it tend to restore it to that course without active intervention on the part of the pilot in the case of an aeroplane, or without active muscular contractions in the case of an animal. Although gliding has probably always preceded flapping in the evolution of flight, stability can also be defined for flapping flight. In this case, there is a continuous series of muscular contractions. We may say that such an animal is stable if the forces acting on it tend to restore it to its course without any modification of that cycle of contractions being required. In practice the most important type of stability is that for rotation about the pitching axis; that is, a horizontal axis normal to the flight path. In both gliding and flapping flight, stability in pitch can be ensured by the presence of an adequate horizontal surface behind the centre of gravity.

The stability defined above is referred to as static stability. The response of a stable aeroplane to a disturbance may be a deadbeat subsidence or an oscillation. Such oscillations will normally be damped, but in rather special circumstances long period oscillations may be divergent. Such divergent oscillations can normally be controlled by a pilot, and it seems unlikely that they are of any great importance in animal flight. The response of an unstable aeroplane to a disturbance is a divergence, whose rapidity depends on the degree of instability.

Flight has been perfected by four animal groups, the birds, bats, pterosaurs, and insects. Too little is known of the post-cranial skeleton of primitive bats for them to be discussed with any certainty. However, it is generally assumed that the bats have been evolved from gliding arboreal mammals functionally similar to the modern cobego *Cynocephalus* (syn. *Galeopithecus*), although there is probably no phylogenetic relationship. In this mammal the patagium forms a web connecting the fore and hind limbs, and extending backward as the interfemoral membrane to include the tip of the long tail. There can be little doubt that it is a stable glider. In the bats the length of the tail, and therefore the size of the interfemoral membrane, is reduced, and the forelimbs are greatly elongated. These changes have the effect of shifting forward the horizontal lifting surfaces relative to the centre of gravity and thus reducing stability.

In the other three groups, there are good reasons to suppose that the earliest forms were stable in the sense defined above. The Archaeornithes possessed a long tail bordered on either side by a row of feathers, the whole forming a very effective stabilizing surface. In the case of the pterosaurs, the earliest known forms from the lower Jurassic belong to the suborder Rhamphorhynchoidea. These forms had a long stiff tail which in at least one genus, *Rhamphorhynchus*, is known to have terminated in a stiffened fluke of skin. This tail must have had a stabilizing function. However, the latest worker on these fossils, Gross[1] believes that the fluke of skin was disposed in a vertical plane. If this is the case, it would have acted as a stabilizer for yawing rotations, that is, rotations about a vertical axis. It would, in fact, be analogous to the fin of an aeroplane rather than to the tailplane. This, if confirmed, is a rather surprising fact, since in an aeroplane, although both pitching and yawing stability are necessary, instability in pitch renders an aeroplane more completely uncontrollable than instability in yaw.

There are also several features of fossil insects from the

Carboniferous to suggest that they were stable. The oldest and most primitive order of winged insects is the Palaeodictyoptera from the lower and middle strata of the upper Carboniferous. They possessed an elongated abdomen, each segment bearing conspicuous lateral lobes, thus forming an effective stabilizing surface. There was also a pair of slender and often greatly elongated cerci. Although such structures would be ineffective as stabilizers on an aeroplane, they are probably quite effective on an insect, due to the greater importance of air viscosity on a small scale (scale in this sense being measured as the product of length and forward speed).

The evolution of instability

It appears, therefore, that primitive flying animals tended to be stable, presumably because in the absence of a highly evolved sensory and nervous system they would have been unable to fly if they were not, just as a pilot cannot control an unstable aeroplane. It is, however, theoretically possible to design an automatic pilot to fly a fundamentally unstable aeroplane. In spite of the obvious practical objections to such a scheme, it would have certain advantages. The first is that an unstable aeroplane could be turned more rapidly. The second advantage lies in the fact that in a stable aeroplane the stabilizing tailplane plays a relatively small part in supporting the weight. In an unstable aeroplane, on the other hand, the elevators would be lowered as the plane flew slower, the tailplane would, therefore, support a larger part of the weight, and thus a lower flying speed could be attained without stalling. (The stalling speed is the minimum speed at which an aeroplane can fly.)

Now it is clear that the practical objections to such a scheme as applied to aeroplanes do not arise in the case of animals. There is, in fact, good evidence that birds do not need to be stable in order to fly. In some birds there is no tail in an aerodynamic sense at all. Other birds, which normally possess a tail, can fly without it; this can often be observed in the case of sparrows which have completely lost their tails. In fact, in most birds the tail does not seem to act as a stabilizer, but as an accessory lifting surface when flying slowly. This can be observed, for example, in the case of gulls. These birds open their tails only when turning sharply or flying slowly. It can then be seen that the slower the bird flies the more the tail is lowered; as mentioned above, this is characteristic of the unstable state.

No such detailed discussion is possible in the case of the pterosaurs, but it is significant that the later upper Jurassic and Cretaceous Pterodactyloidea completely lacked a tail.

In the case of insects, it is impossible to make any generalizations, since there is such a wide adaptive radiation within the group. It is probable that some orders, for example, the Ephemeroptera, are stable. However, in the case of the Diptera the work of Hollick[2] and Pringle[3] has shown the importance of the arista and halteres during flight. Indeed, in the case of the Diptera, so far from being stable, the forces acting on a fly are not even in equilibrium in the absence of sensory input from the arista and muscular response to this input.

To a flying animal there are great advantages to be gained by instability. The greater manoeuvrability is of equal importance to an animal which catches its food in the air and to the animals upon which it preys. A low stalling speed is important in a number of ways, and particularly to larger animals. For a set of geometrically similar animals, the stalling speed increases as the square root of the linear dimensions. Therefore a successful landing may be possible in the case of a large animal only if it can reduce its stalling speed, and instability is one of the ways in which this may be done. The account given above of the way in which gulls use their tails illustrates this point. Thus it is possible that the evolution of a pterosaur as large as *Pteranodon* depended on the prior evolution of instability. In extreme cases a lower stalling speed may make hovering flight possible.

It is also important to realize that we are not concerned with a change from stability to instability which must be made in a single step. Any reduction in the degree of stability will be an advantage provided there is a parallel increase in the efficiency of control. This can be seen by analogy with aeroplanes. Transport aeroplanes are normally designed with a fairly high degree of stability, since safety in steady flight is of greater importance than manoeuvrability. In fighter aircraft, however, manoeuvrability is of first importance, and the stability margin is usually reduced to a minimum. It is, therefore, possible to see how instability may have been evolved gradually.

Palaeontologists will have to solve the question of the relative times taken to evolve stable flight, with the relatively coarse controls needed for it, from walking and climbing; and of unstable from stable flight. Unfortunately we only know Archaeornithes from one horizon; on the other hand the Rhamphorhynchoidea persisted for a time measured in tens of millions of years, as did the stable Paleaodictyoptera. It is possible that the evolutionary changes needed for stable flight could be made rather quickly, while the nervous and sensory adjustments needed for unstable flight were inevitably slower. If, as is also possible, the bats evolved rather quickly to instability, this may be due to the greater adaptability of the mammalian brain.

If the conclusions of this paper are accepted the study of the remains of primitive flying animals, and even experimental studies on full-scale models of them, will acquire a special importance as throwing new light on the functional evolution of nervous systems.

There are, of course, several other cases where similar deductions can be made as to the evolution of systems of which no direct fossil evidence exists. Among the most obvious is the need for a highly developed vasomotor system in large land animals which change their posture. A dinosaur standing on its hind legs without the previous evolution of such a system would have suffered from cerebral anaemia. However, it is doubtful whether the palaeontological evidence for such evolution is as clear in any other case as in that of flight.

Summary and conclusions

Stability in gliding and flapping flight is defined. It is argued that the earliest birds, pterosaurs, and flying insects were stable in the sense defined. This is believed to be because in the absence of a highly evolved sensory and nervous system they would have been unable to fly if they were not.

The advantages of instability to a flying animal are discussed. It appears that in the birds and at least some insects, and probably in the later pterosaurs, the evolution of the sensory and nervous systems rendered the stability found in earlier forms no longer necessary.

Acknowledgements

The writer wishes to asknowledge the assistance he has received from discussing this paper with Professor J. B. S. Haldane and Mr L. G. Morris.

REFERENCES

1. W. Gross (1937) *Abh. Senckenberg. Naturf. Ges.*, 437, 1–16.
2. F. S. J. Hollick (1940) *Phil. Trans. roy. Soc. B*, 230, 357–90.
3. J. W. S. Pringle (1948) *Phil. Trans. roy. Soc. B*, 233, 347–84.

Evolution and History

Sociologists tend to be suspicious of the application of biological ideas to their subject. So much nonsense has been written in the guise of 'Social Darwinism', and so many crimes justified by theories of racial superiority, that this suspicion is perhaps justified. Therefore I would like to stress at the outset what appear to me to be the limitations of biological thinking in the social sciences. I do not think that biological evolution has itself been an important motive force for change during human history, although it was responsible for the origin of those specifically human characteristics which made history possible. Doubtless the human species has continued to evolve throughout historical times, but at a rate which must have been very slow compared to the rates of historical change, mediated by cultural rather than by genetic transmission from generation to generation.

Nor do I think that the cultural differences between social classes, or between nations or races, are to any significant extent due to genetic causes. Social classes are seldom reproductively isolated from one another to an extent sufficient to permit any significant genetic difference to arise between them. This is not true of human races, which have in the past been sufficiently isolated for recognizable and genetically determined physical differences to arise, although physically distinct populations are usually connected by others of intermediate type. It is therefore conceivable that genetically determined differences in emotional and intellectual capacities may also exist. One reason for doubting that they do so is that the relative levels of cultural and technical achievement of different populations are not always the same at different periods of history. A genetic interpretation of history would require us to suppose that a thousand years ago the Arabs were genetically better fitted for scientific inquiry than the inhabitants of Western Europe, whereas today the reverse is true.

It is more sensible to suppose that the factors which influence the cultural achievements of a population are not to any great extent genetic.

In other words, genetically determined differences between human populations at different times or in different places can probably safely be ignored when considering the history of the last ten thousand years, although genetically determined differences between the individuals composing a society at any one time can not.

The two uses of analogy

The main reason for applying evolutionary ideas to history lies in the hope of drawing helpful analogies between the two processes. Before considering some of the analogies which can be drawn, I would like to discuss in general terms the uses of analogy in science. Consider first a simple and well-understood analogy. Figure 1 shows two systems, one mechanical and the other electrical, which have in common the property of oscillating harmonically according to the equation $d^2x/dt^2 + Cx = 0$, where the constant C depends in the one case on the stiffness of the spring and the mass of the weight, and in the other on the capacitance of the condenser and the inductance of the coil. The analogy between the two systems consists in the similarity in their behaviour, which in turn depends on the similarity in the relationships between their parts. It does not in any

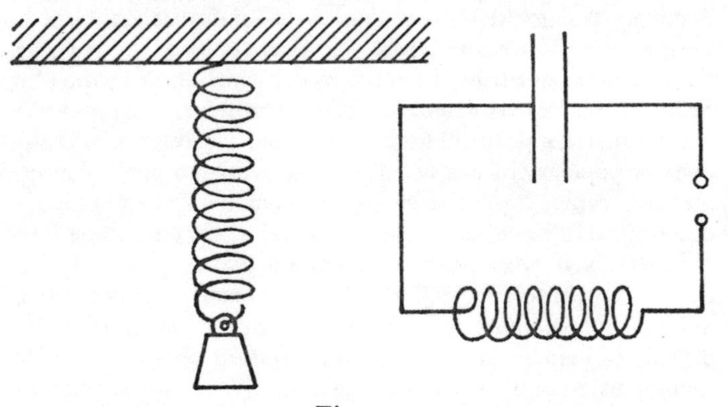

Figure 1.
The analogy between mechanical and electrical oscillating systems

way depend on a similarity between mass and inductance as such. Two machines such as these, whose behaviour is identical, are said to be isomorphic.

There are two possible uses of such an analogy. First, if the analogy is exact it can be used for predicting the behaviour of one system by observing that of another. For example, since it is cheaper to build and to modify electrical circuits than complex structures, it was at one time the practice during the design stage to predict the natural modes of vibration of an aeroplane's wing by building an analogous electrical circuit; such a circuit was in fact an analogue computer. Today general-purpose analogue computers have been designed which, by alteration of their internal connections, can be made to solve a wide variety of problems. There seems little prospect that analogies between evolution and history will ever have a predictive value in this way. Even if it is possible to write down the initial conditions and mathematical equations which describe an historical situation with sufficient accuracy for prediction to be possible, it will always be cheaper to use an electrical than a biological computer to solve them. Unhappily, I cannot look forward to the day when the course of a trade cycle will be predicted by using populations of fruitflies or of bacteria.

But analogies have another use; they may help us to think about unfamiliar things. To return to figure 1, all human beings acquire during their childhood an understanding of mechanical phenomena, since they possess mechanical and visual receptors. Consequently, when first confronted with electrical phenomena, they find them easier to comprehend if they can draw analogies with more familiar processes. Most of us would, I think, admit that we found direct current circuits easier to understand by recognizing the analogy between electrical potential and head of water. Perhaps there are some of you whose minds are sufficiently adept at abstraction to find such analogies unnecessary and therefore misleading, but I know it is not so in my case.

Now the use of analogies in this way is widespread in biology. It is not necessary that two systems be isomorphic, or that an exact mathematical description be given of either of them, provided that they have something in common in their behaviour. Examples of such analogies are the 'psycho-hydraulic' model of the brain evolved by Lorenz, and the comparison by Waddington of a developing organism with a ball rolling down an 'epigenetic landscape'. I do not know whether the former analogy has been useful, but I have during the last few years performed experiments which would probably not have occurred to me had I not been familiar with Waddington's model.

That analogies can readily be drawn between biological and social phenomena is apparent in such phrases as 'the head of state', 'arterial road' and 'the arm of the law'. If such analogies can be made more precise they may give us new ideas, although they cannot in the nature of things prove that those ideas are correct. But it seems to me that what we need above all else in the study of society is theories which we can test. There is nothing easier, either in biology or in sociology, than to collect facts not previously known. Almost any facts about human beings have an intrinsic interest for us, but the mere collection of facts, however interesting or however true, does not constitute science. It is an essential feature of the scientific method that we should put forward theories or hypotheses which are in principle capable of being contradicted by observation, and that we should then perform experiments or collect facts to see whether they do or do not contradict our theories. If the drawing of analogies can help us to formulate such theories, then it is justified. Some of the analogies suggested in Professor Waddington's essay[1] seem to be of this nature. For example, he suggests that certain groups of ideas or customs may be transmitted together from one culture to another because they happen to occur together in the 'donor' culture, rather as genes may be linked on the same chromosome, and not because there is any necessary connection between the ideas and customs as such. This is a suggestion which is capable of disproof, and therefore informative. Does what a biologist would call the 'host' culture accept groups of customs merely because they are found together in the donor, or is it in fact selective, adopting only those customs relevant to its own conditions?

But although analogies between historical and biological processes can readily be drawn, I am doubtful whether any general analogy between them is likely to be helpful. In both, we are confronted by processes of continuous rather than of purely cyclical change. Also the behaviour of both systems depends on the interrelationships of entities which are not only very numerous (as they are in a crystal or in a perfect gas), but also of many qualitatively distinct kinds. Consequently the behaviour of both systems is of great complexity, so that theory cannot in general hope to predict the long-term behaviour of a system, but only the immediate effects of some interference with it. But when these general resemblances have been noted, I do not think that the two systems have much else in common. In particular, the ways in which the various entities concerned are interrelated are quite different in the two cases, as will become apparent if our current theory of evolution is presented in diagrammatic form.

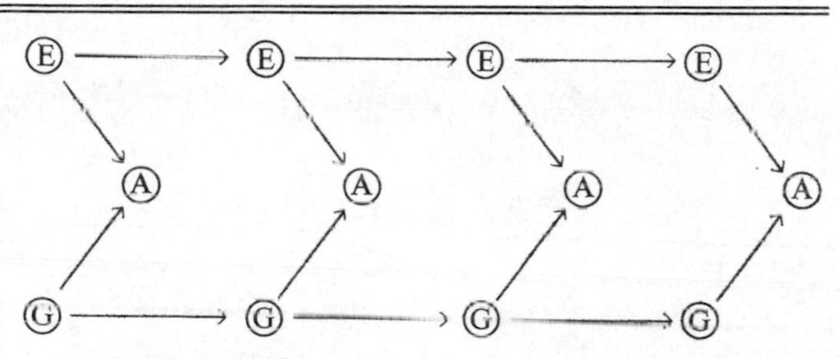

Figure 2. Diagram of the theory of heredity

In figure 2, G represents the fertilized egg, A the adult individual which develops, and E the environment in which it does so. The arrows from G to G represent Weissman's theory of the continuity of the germ plasm, those from G to A the process of development, and those from E to A the fact that the kind of adult which develops depends on the environment. The absence of an arrow from A to G is commonly expressed by saying that acquired characters are not inherited, and the absence of an arrow from A to E implies that animals do not by their own activities alter their environment. This is of course not strictly true; but it is true that the influence of a population on its environment is not usually of major importance either as a conservative mechanism or as a cause of evolutionary change, although it is often of vital importance to an understanding of the ecology as opposed to the evolution of a population. Figure 2 does not indicate the causes of evolutionary change. This is done in oversimplified form in figure 3, in which the diverging arrows from each G represent processes of segregation and mutation, and natural selection has been incorporated by showing that the environment, although it cannot alter the germ plasm adaptively, can destroy ill-adapted adults, and so prevent the transmission of certain kinds of germ plasm.

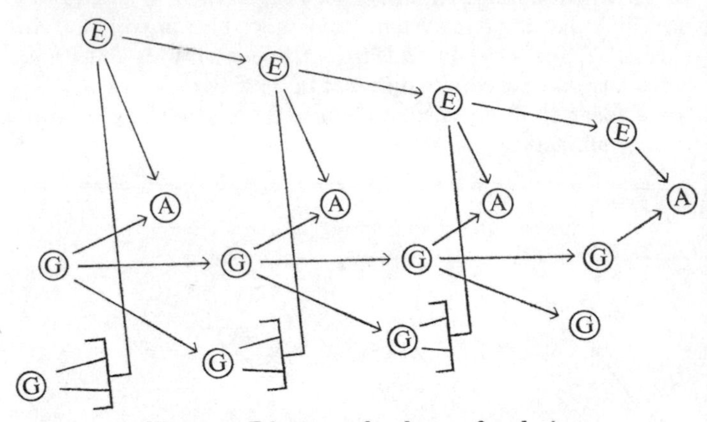

Figure 3. Diagram of a theory of evolution

The most general kind of statement which can be made about history is indicated in figure 4, in which C represents the customs and ideas of a people at a given time, and E their environment, including such things as the houses in which they live, the tools they use, and the animals and plants which they have domesticated. The arrows connecting E to C indicate that people's ideas are influenced by their circumstances, and those connecting C to E that man modifies his own environment to an extent which cannot be ignored, as can (at least for some purposes) the comparable process in animals. Now figure 2 is far more informative than figure 4; it is informative because of the arrows it leaves out. Figure 4 permits any present circumstance to influence any future one; since it excludes nothing it predicts nothing. Theories of history[2] in effect state either that some of the arrows in figure 4 are more important than others, or apply restricted meanings to C and E, suggesting that some particular aspects of human behaviour and environment are of major importance in determining the course of history.[1]

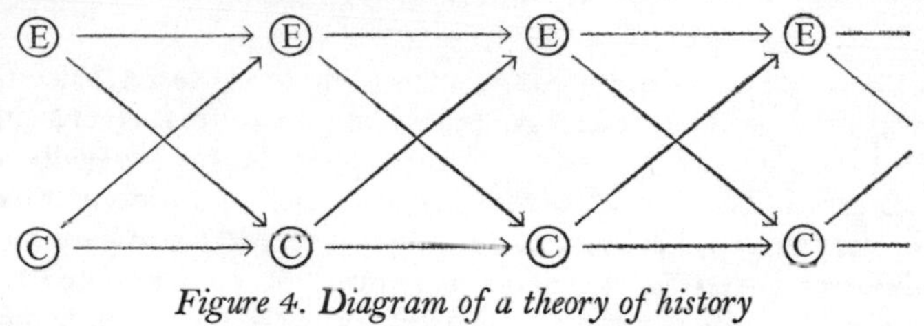

Figure 4. Diagram of a theory of history

For example, figure 5 shows two views of history which are formally similar to figure 2; if either were true, useful analogies might be drawn between history and evolution. Figure *5a* represents an idealist theory of history; what is important is the history of ideas, which develop according to their own logic, but which also determine the kind of world in which men live. It is not necessary to accept the Marxist thesis that man's being determines his consciousness to feel that this view leaves out too much to be helpful. Even a subject such as mathematics, which is capable of considerable independent development in directions determined by its own internal logic, has been greatly influenced by problems of practice. Figure *5b* represents a theory of 'economic determinism', which to the best of my knowledge has never been held by anyone, although it has frequently been fathered on Marx by his opponents.

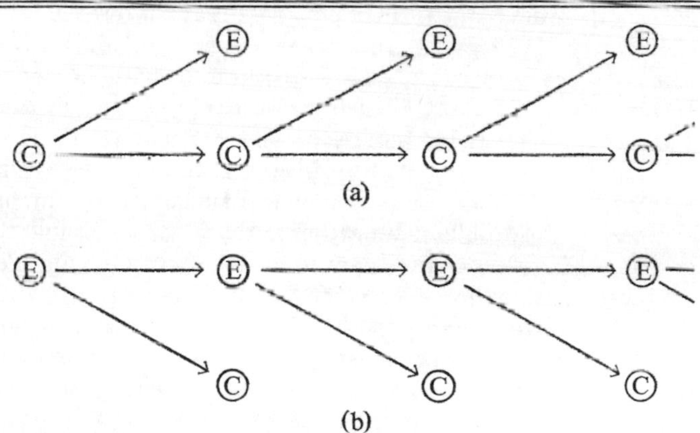

Figure 5. Diagrams of theories of history formally similar to the theory of heredity

It is worth considering Marx's theory in more detail, since it was developed to act as a guide in changing society, and has in fact been widely applied for that purpose. It was avowedly a dialectical theory, and so does not lend itself to representation by diagrams which in effect describe the relationships between different parts of a machine. However, Marx attempted to develop an informative theory from figure 4, not by omitting some of the arrows, but by applying restricted meanings to the terms C and E. C represents the relationships between men in the process of production, and E the tools used in production. Marx suggested that the course of history was determined by the evolution of this sub-system, which he called the means of production, and which in turn influenced ideas of religion, philosophy, art, and politics. Further, he suggested that the system represented by C, the social relations of production, show a greater conservatism than do the techniques of production. Consequently the social relationships come in time to act as barriers to the further development of production, and are then changed in a revolutionary manner. At such periods, political and other ideas react back on his basic system, and play a part in the transformation of society. This is perhaps not a fair picture of Marx's views; but in the present context the important point is that his theory is more nearly represented by a diagram of the form of figure 4 than by figure *5a*.

Although few western sociologists would call themselves Marxists, many of Marx's ideas are in fact tacitly accepted, particularly in the study of earlier historical periods. My own view is that his was the most successful attempt yet made to develop an informative theory of history, and that the weakness of Marxism lies not in any fundamental error in his approach to the problem, but rather in the failure of his followers to treat his ideas critically, or to modify them in the light of advances in other fields, as Darwinism has been modified in the light of Mendelian genetics. Now if it is true that any adequate theory of history must take into account all the causal connections indicated in figure 4 (and so can become informative only by singling out from the totality of events certain processes which dominate the course of history, either as Marx did or in some other way), then it follows that no formal analogy between historical and evolutionary processes as a whole is possible. In other words, I do not think that a helpful theory of history can be derived by starting from evolution theory, and attempting to find historical analogues for the various entities and processes involved.

The problem of constancy

Even so, it may still be possible to draw helpful analogies between particular aspects of the two processes. There is in fact one particular habit of thought characteristic of biologists which might usefully be borrowed by historians. Biologists are continuously interested in mechanisms which preserve constancy, of faunas, of species, of individuals, or of kinds of molecules. It would be interesting to know how far comparable mechanisms are responsible for preserving the constancy of human institutions and patterns of behaviour and belief. The constancy of human cultural patterns during the palaeolithic was extreme; for example, Acheulian 'hand axes' were manufactured with relatively little change for some 100,000 years over a large part of Europe, Africa, and Asia. Such stability calls for an explanation. Or to take a more recent example, Gibbon once remarked that the curious thing about the Roman Empire was not that it declined but that it survived for so long. It may be that the search for the causes of constancy in human affairs may prove as fruitful as has the comparable study of homeostasis in biology.

In biology, we know of two rather different kinds of process for ensuring constancy, which I will call 'copying plus selection' and 'error regulation'. The conservation of heredity appears to depend on the exact copying of DNA molecules, as a scribe might copy a manuscript. Since no copying process can be completely free of errors, this process would lead slowly but inevitably to chaos were there not some mechanism for eliminating the mistakes which are made. An error, called a mutation, is not corrected once made; it is eliminated by natural selection, or occasionally incorporated by selection in place of the original. In contrast, the physiological constancy of individuals is maintained by error regulation; if you get too hot you sweat, if too cold you shiver. The regulation of development appears to work in a similar manner, except that what is maintained is not a steady state but a particular pattern of change, which Waddington has called a chreod.

It is easy to point to error-regulated constancies in society.

The relative prices of commodities remain approximately constant through the action of the law of supply and demand. Or to take another example, it appears to be a feature of the present political systems of Britain and the United States of America that there should be two major political parties, of approximately equal strength, succeeding one another in office. I do not suggest that this is one of the eternal verities, but the balance has been maintained over an appreciable period. To explain this persistence, it would be necessary to show why it is that in time the party out of power gains support at the expense of the party in power.

The second type of mechanism, that of copying plus selection, seems to have no close analogue in human society. It is true that rote-learning is a copying process. But errors are not corrected by a selective process, unless we regard the failing of unsuccessful candidates as a form of natural selection. But it seems more natural to suppose that if errors of transmission occur, they are corrected so as to make sense, although not necessarily the original sense; to the best of our knowledge genetic mutations cannot be corrected so as to make biological sense.

Limitations of analogical method

In conclusion, I would like to return to the comparison between Marx and Darwin. Darwin put forward a theory to explain how evolution had occurred. Marx put forward a comparable theory of history, but he claimed to have done more than this; he derived from Hegel categories of change which he held to be characteristic not only of history but of all natural systems, whether social, biological or physical. In the present context, this amounts to the claim that what history and evolution have in common is that both obey dialectical laws. Now there is no *a priori* reason why this claim should not be true, or, if true, why it could not be substantiated, as Engels attempted to do in *The Dialectics of Nature*. For example, many simple systems, mechanical, electrical and hydrodynamic, have the property of vibrating harmonically, and this category of behaviour could be recognized without making a detailed analysis of any particular case. Marx in effect claimed that dialectical categories of behaviour could be recognized as characteristic of more complex systems.

In a limited sense, most scientists would agree; for example, few use the phrase 'transformation of quantity into quality', but everyone is aware of the dangers of extrapolation. But is the dialectical approach of any more general value in science, and in particular in biology? I find this a very difficult question to answer. Often it seems to make simple problems complicated, and to substitute vagueness for precision. Occasionally it can have disastrous consequences. As will be clear from a comparison of figures 2 and 4, evolution would be more like history if Lamarckian inheritance were true; if, in figure 2, there were an arrow passing from A to G. This seems to have led some Russian biologists to defend what I believe to be an erroneous theory of genetics by the argument that it must be true because it is Marxist. Such an argument is a misuse of analogical thinking. If a scientist is convinced of the dialectical nature of history, it is natural that this should lead him to seek for comparable phenomena in biology, and perhaps to be predisposed in favour of a Lamarckian theory of genetics. But this in no way constitutes evidence in favour of such a theory, which can come only from observation and experiment, a point which was recognized by some but unfortunately not by all the participants in the Russian controversy. Analogies, Marxist or otherwise, may be helpful in suggesting theories, but are irrelevant when it is a question of confirming or disproving them.

There is however one very simple reason why a study of history may be helpful in suggesting ideas to biologists. Although historians may not yet have achieved an agreed theory by which to work, they know incomparably more about the actual course of history than biologists do about the course of evolution. Perhaps as a consequence, historians are obliged to recognize the dialectical nature of many historical processes, whatever words they may use to describe them, whereas most evolutionists are wedded to the 'inevitability of gradualness', a view which may be largely correct, and which in any case was held by Darwin. Perhaps the writer on evolution whose views would most appeal to a dialectical materialist was Richard Goldschmidt,[3] in particular in his views on systemic mutation, on the relation between organism and environment, and on the integration of the genetic material of a chromosome. There are features of Goldschmidt's views on systemic mutation which appear to me untenable. But he may have been right in thinking that evolution is a more dialectical and less Fabian process than many of us have supposed.

Notes and References

1 C. H. Waddington (1961) The human evolutionary system. *Darwinism and the Study of Society*, pp. 63–81 (ed. Banton, M.P.). London: Tavistock.

2 I omit what is perhaps the most popular 'theory', that is, that no theory of history is possible: history is certainly very complicated, but that is no excuse for throwing up the sponge; evolution is very complicated too.

3 R. Goldschmidt (1940) *The Material Basis of Evolution*. Newhaven, Connecticut: Yale University Press.

The Arrangement of Bristles in Drosophila

This paper was written with K. C. Sondhi, when both authors were at the Department of Zoology, University College, London. It is dedicated to Professor L. C. Dunn in recognition of his long and distinguished career.

Much of the geometrical complexity of animals and plants arises by the repetition of similar structures, often in a pattern which is constant for a species. In an earlier paper[1] some of the mechanisms whereby a constant number of structures in a linear series might arise were discussed. In this paper an attempt is made to extend the argument to cases where such structures are arranged in two-dimensional patterns on a surface, using the arrangement of bristles in *Drosophila* as illustrative material.

The bristles of *Drosophila* fall into two main classes, the microchaetes and the macrochaetes. A bristle of either type, together with its associated sensory nerve-cell, arises by the division of a single hypodermal cell. The macrochaetes are larger, and constant in number and position in a species, and in most cases throughout the family Drosophilidae. The microchaetes are smaller and more numerous, and show no fixed number or pattern in a species, although they do show some regularity in spacing. A number of mutants are known which alter the number and arrangement of the macrochaetes, usually by eliminating one or more pairs.

An explanation is put forward of the variations in the number and arrangement of the microchaetes, and of the macrochaetes in some mutant stocks, in terms of a common morphogenetic model. This model is an extension of that suggested by Stern[2] in the light of mathematical considerations due to Turing.[3] It is hoped that this model may have some relevance to the arrangement of other repeated structures.

Stern[2] has analyzed a number of bristle patterns in *Drosophila* by means of genetic mosaics. In the mutant achaete, which removes the posterior pair of dorsocentral bristles on the thorax, he showed that if in a predominantly wild-type fly a patch of genetically achaete tissue covered the site of the bristle, no bristle was formed; but if in a predominantly achaete fly the site of the bristle was covered by genetically wild-type tissue, a bristle was formed. He suggested that the presence of a bristle required the existence of a 'prepattern' determining its position, and the existence of cells competent to respond to the prepattern by forming a bristle. The normal prepattern exists both in wild-type and in achaete flies, but genetically achaete cells are incompetent to respond to this prepattern. An essentially similar conclusion was reached by Maynard Smith and Sondhi[4] from a study of populations of *D. subobscura* homozygous for the mutant ocelli-less, which were selected for different expressions of the mutant.

The prepattern is most easily pictured as the distribution of an inducing substance with regions of high and low concentration, the regions of high concentration occurring at sites where bristles later form. A process whereby such a distribution could arise has been suggested by Turing.[3] He considered the distribution in a morphogenetic field of two chemical substances, or 'morphogens', together with an adequate supply of substrate from which they could be synthesized. These morphogens are free to diffuse and to react with each other. He showed that for certain values of the rates of reaction and diffusion the initial homogeneous equilibrium was unstable; any disturbance of the equilibrium, for example by Brownian movement, leads to the development of a standing wave of concentration of the morphogens. The actual pattern of peaks and valleys of concentration depends on the size and shape of the field, and on the 'chemical wavelength', that is, the preferred spacing between peaks, which in turn depends on the rates of reaction and diffusion; it does not depend on the nature of the initial disturbance.

This provides a simple model of the process whereby a prepattern could arise. In cases in which a pattern is constant throughout a species, it is a more satisfactory one than Wigglesworth's competitive model.[5] According to Wigglesworth's model, the positions of the bristles will depend in part on which particular hypodermal cells happen by chance to be the first to differentiate, whereas Turing's mechanism would give rise to a pattern independent of the initial chance disturbance. Thus a competitive mechanism could explain the arrangement of a series of structures whose only regularity is the approximately equal spacing between them, but not of structures whose arrangement is constant from individual to individual.

The way in which the arrangement of structures may depend on the shape of the field as a whole has been shown by Sengel's work[6] on the development of feather papillae on the skin of the chick *in vitro*. If skin is removed from the dorsal region of the embryo when the feather rudiments have just become visible, these rudiments disappear, and later new feather papillae develop in different positions, the first to appear forming a row along the centre of the explant.

Turing's two-morphogen model is, however, too simple to explain all the facts. As Waddington has pointed out,[7] it would predict that different patterns would arise if particular stages of differentiation occurred in embryos of different sizes, and this is not always the case. But in spite of this and other difficulties, Turing's suggestion of how a prepattern might arise is probably along the right lines.

The arrangement of microchaetes

For reasons of economy, it would be desirable to explain the arrangement of microchaetes and of macrochaetes by similar mechanisms, differing only in the accuracy with which they are regulated. Fortunately, there is some observational evidence that similar mechanisms are involved. In this section it will be shown that the arrangement of microchaetes does show traces of the kind of regularity to be expected if their positions depend on the shape of the field as a whole; in the next section it will be shown that at least some abnormalities of macrochaete arrangement are of the kind to be expected from variations of a prepattern arising in the way suggested by Turing.

Figures 1*a* and 1*b* show two solutions of Turing's equations in a uniform rectangular field, the dots corresponding to peaks of concentration of one of the morphogens; the solutions differ only because slightly different reaction rates have been assumed. Figures 1*c*, *d*, and *e* show the arrangement of microchaetes on the sternites of the fourth abdominal segment of three individuals of *D. subobscura.* The first shows a somewhat irregular pattern, the second shows clear rows of bristles parallel to the boundaries of the sternite, and the third shows diagonal rows. The resemblance between the two latter sternites and the theoretical distributions is obvious. Sternites with these regular patterns are quite common in flies with low numbers of bristles, but irregular arrangements are more usual in flies with larger numbers of bristles. Even an occasional sternite of the type shown in figure 1*d* is sufficient to show that the pattern depends on the field as a whole, and such sternites are by no means uncommon.

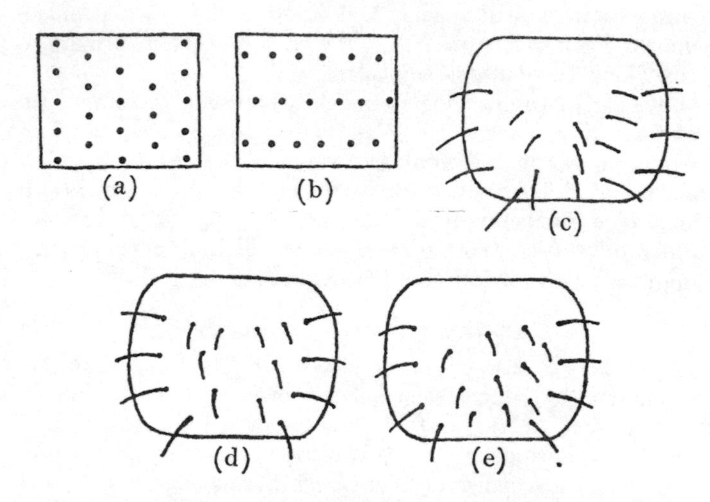

Figure 1. a and b. Solutions of Turing's equations in two dimensions; the dots represent peaks of concentration of a morphogen. c, d and e. The arrangement of microchaetes on the fourth abdominal sternite in three individuals of D. subobscura.

Macrochaetes

Figure 2*a* shows the arrangement of macrochaetes and ocelli on the top of the head of *Drosophila.* The sex-linked recessive mutant ocelli-less in *D. subobscura* removes most of these structures. There is however considerable variation in

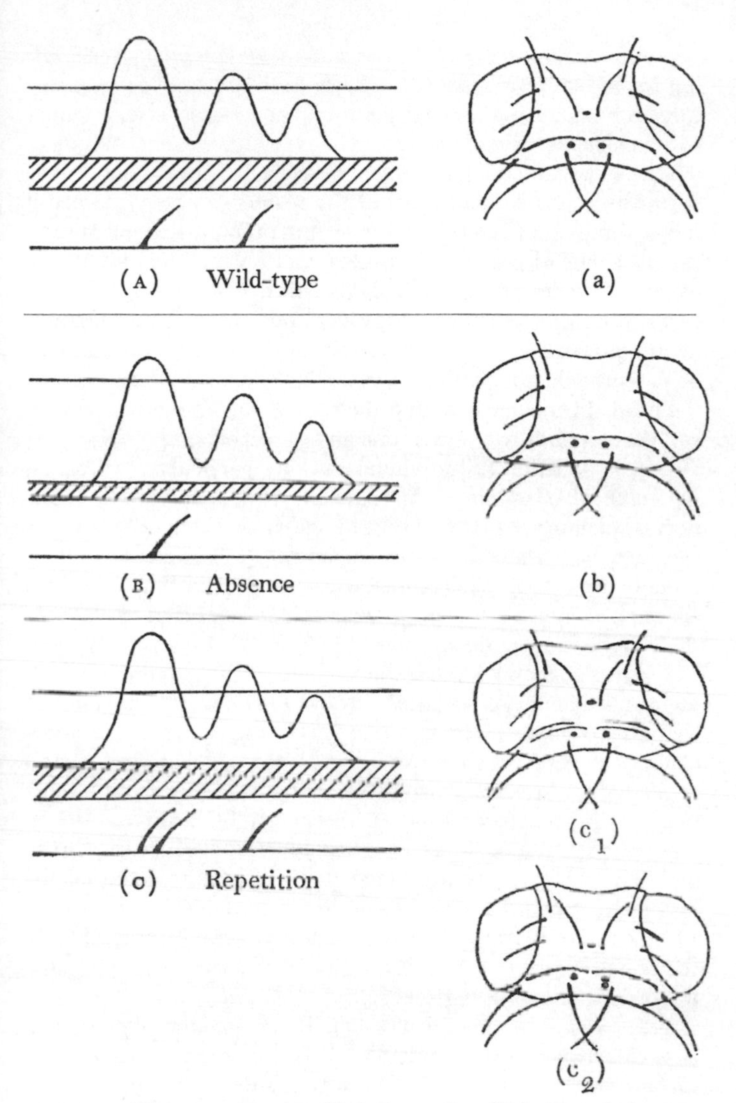

Figure 2. The arrangement of bristles and ocelli in D. subobscura *in a, the wild-type; and b, c_1, and c_2, individuals homozygous for the mutant ocelli-less. In the diagrams on the left, the curved lines represent the prepattern, pictured as a varying concentration of an inducing substance; the hatched areas the concentration of precursor; and the upper horizontal lines the threshold level which the prepattern must reach if it is to induce a structure*

populations homozygous for the mutant, and by selective breeding for individuals with a larger or smaller number of structures, or with only particular structures present, a wide range of phenotypes, including the wild-type, have been obtained. These experiments have been described elsewhere.[1, 8, 9, 10] It has been shown that most of the results can be explained if it is assumed that there is an unvarying prepattern determining the positions of ocelli and bristles, and a varying amount of a common 'precursor' of ocelli and bristles. The absence of structures in unselected mutant stocks is due to the small amount of this precursor, but selection can both increase the amount of precursor, and also concentrate it in particular regions of the head. In this explanation the concept of 'amount of precursor' corresponds to Stern's concept of level of competence.

But, in addition to individuals lacking particular structures, we have also obtained individuals with various other abnormal phenotypes. It will now be shown that these phenotypes can also be explained in terms of the prepattern-precursor model. Reference will be made to the populations in which these phenotypes occur only when this is helpful in explaining their origin during development.

Figures 2, 3 and 4 show the various possible ways in which abnormal phenotypes can arise, and examples of phenotypes thought to have arisen in these ways. Figures 2 and 3 show changes arising from differences in the amount and distribution of precursor, and figure 4 changes arising from alterations in the prepattern. The various possibilities are as follows.

Absence of structures. Due to a low level of precursor (figure 2*b*). These are the typical abnormalities in unselected ocelliless populations. Figure 2*b* shows the commonest phenotype in a population which had been selected so as to concentrate the precursor in the posterior region of the head and to remove it from the anterior region.

Repetition of structures (figure 2*c*). If the amount of precursor is greater than in the wild-type, it is possible that two structures should develop in response to a single peak of the prepattern. Figures 2c_1 and c_2 show repetition of bristles and of ocelli respectively. Such repetition occurs mainly in populations which have been selected for an increased number of structures, and therefore presumably for an increased amount of precursor.

Neomorphs (figure 3*d*). If there can exist in mutant flies peaks of the prepattern to which genetically mutant tissue is incompetent to respond, it is possible that there also exist 'submerged' peaks of the prepattern to which wild-type tissue does not respond. If so, bristles might appear at the sites of these submerged peaks in mutant populations selected for

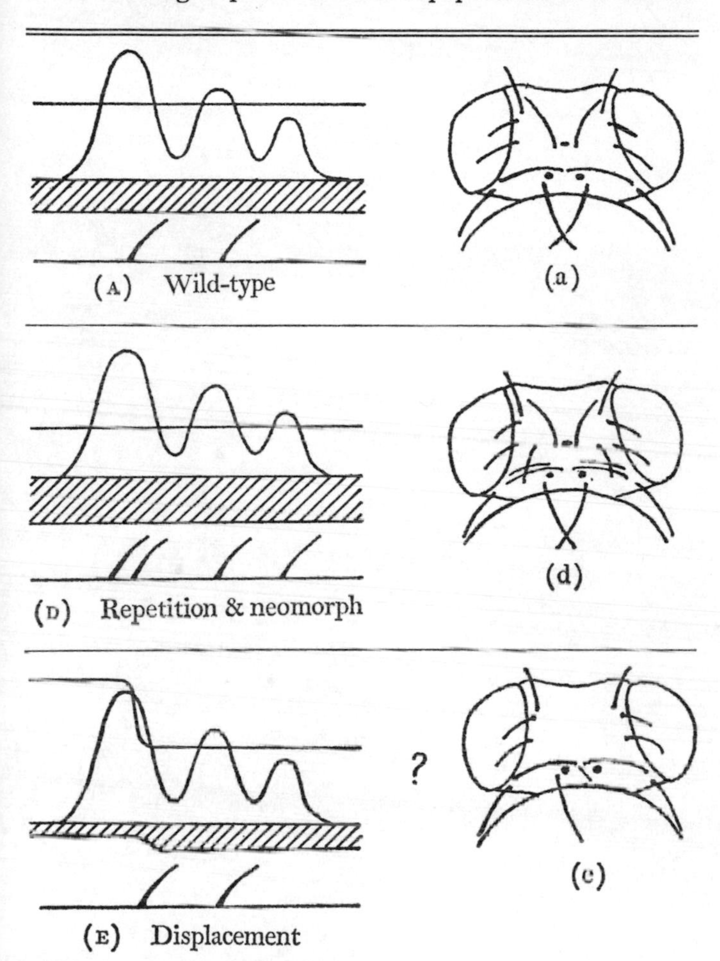

Figure 3. The arrangement of bristles and ocelli in D. subobscura *in a, the wild-type; and d and e individuals homozygous for the mutant ocelli-less.* See *legend to figure 2*

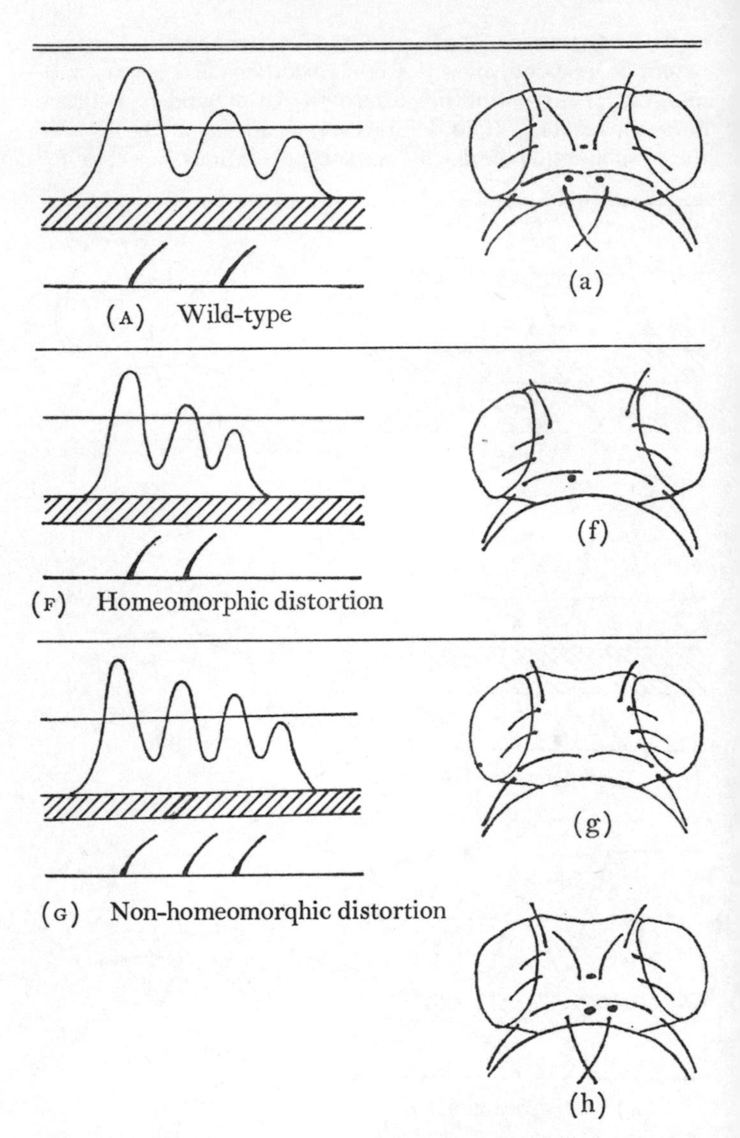

Figure 4. The arrangement of bristles and ocelli in D. subobscura *in a, the wild-type; and f, h, g, individuals homozygous for the mutant ocelli-less.* See *legend to figure* 2

increased competence. Figure *3d* hows a phenotype thought to arise in this way. The new bristle occurs in about 5 per cent of flies in a population which has been selected for many generations for an increased number of structures. It is confined to this population, and occurs only in individuals in which all the normal structures are present. The bristle is constant in its position and orientation, and closely resembles in these respects a bristle which occurs typically in flies of a related family.[10] The main reason for thinking that it arises by the process shown in figure *3d*, and not from a change in the prepattern, is that its presence is not associated with any change in the positions of other bristles.

Displacement (figure *3e*). If the precursor is absent at the peak of the prepattern, but present a short distance away, it is possible that a bristle would develop in a position slightly displaced from the normal. Stern[2] has plausibly interpreted cases of displacement in his material in this way, but in our material we cannot always distinguish displacements arising in this way from those due to a distortion of the prepattern. This difficulty arises if a bristle is displaced without any associated displacement of neighbouring structures. This happens most commonly in the case of the postvertical bristles, particularly in populations selected for a low number of structures. The base of the bristle is displaced posteriorly and medially, and the displaced bristle is directed in an antero-medial direction (figure *3e*). Such displacements could be due to either of the mechanisms shown in figures *3e* or *4f*, although in the case of the postverticals the latter seems the more likely explanation, since the displacement is almost always in the same direction.

Homeomorphic distortion. That is, a prepattern with the same number of peaks, but of a different shape (figure *4f*). Figure *4f* shows a phenotype thought to arise in this way; it is common in populations selected for a low number of structures on the centre of the head. The displacement of a single bristle could also be due to the mechanism shown in figure *3e* above, but in these flies all three orbital bristles are displaced posteriorly, and this could happen only if the prepattern itself is distorted.

Non-homeomorphic distortion. That is, a prepattern with a different number of peaks from the wild-type (figure *4g*). Figure *4g* shows a phenotype thought to arise in this way;

it also is common in populations selected for a low number of structures on the centre of the head. Four orbital bristles are present instead of three, and usually only the most anterior one is exactly in the position occupied by a bristle in the wild-type. The four bristles are evenly spaced; we have never observed the repetition of an orbital bristle, comparable to the repetition of a vertical bristle shown in figure 2*c*1. The interest of this particular phenotype is that it shows an unusual change in the arrangement of the macrochaetes, but one which arises by a mechanism which we believe to be that typically responsible for variations in the arrangement of microchaetes. In terms of Turing's model, the ratio between the chemical wavelength and the size of the field in which the waves are developing can vary within certain limits without involving any change in the actual pattern formed; but ultimately a threshold would be reached, involving the appearance of an additional bristle or bristles, and a respacing of other bristles in the field.

These mechanisms account satisfactorily for all common abnormalities observed in ocelli-less populations. But occasionally there occur additional ocelli or bristles, at sites which are not occupied by such structures in the wild-type, and which vary from individual to individual in an irregular manner. Figure 4*h* shows such an additional ocellus. Additional ocelli and bristles of this sporadic type occur in populations selected for a high number of structures, with frequencies of about 0·9 per cent and 0·6 per cent respectively.

There is one unexpected feature of these results. It appears that a mutant, which was at first thought to affect only the amount and distribution of the precursor, also, although less commonly, modifies the prepattern, since the phenotypes in figures 4*f* and *g* cannot easily be explained without this assumption. The mutant is therefore pleiotropic, in that it modifies two separate morphogenetic processes. But it seems plausible to suggest that the primary effect of the gene is to alter the concentration of precursor, and that this in turn may, in extreme cases, modify the development of the prepattern.

Discussion

The development of specific structures at specific sites in *Drosophila* has been regarded as the result of two processes, one concerned with the formation of a prepattern which determines the positions at which structures are formed, and the other responsible for the competence of cells to respond to this prepattern by forming the appropriate structures. The justification of this division is that the two processes can vary independently of one another. Our strongest evidence for thinking that the competence can vary while the prepattern remains unchanged comes from an ocelli-less population in which individuals were chosen as parents if they had the two posterior ocelli but lacked the anterior one.[4] In this population the frequency of the selected phenotype increased from 15 to 64 per cent. At the same time, the frequency of individuals possessing the ocellar bristles, which lie close to the anterior ocellus, decreased almost to zero. In those individuals which did possess the anterior ocellus, the ocellus was not displaced posteriorly, but was usually much smaller than in the wild-type. These results only make sense if it is supposed that the prepattern determining the position of the anterior ocellus was unaffected by selection, and that the population changed because the competence of the cells to respond was reduced in the anterior part of the head. In contrast, our strongest reasons for thinking that the prepattern can change are the phenotypes shown in figures 4*f* and *g*.

The distinction between prepattern and competence is therefore made necessary by the nature of the variation observed. Variations in adult structure do not necessarily, or even usually, reflect changes in prepatterns. Differences between individuals may arise because of genetically determined differences in competence between their cells. The importance of this distinction has been increased by the work of Kroeger[11] on the wing-hinges of *Ephestia*. He has been able to show that differences between serially homologous parts of the same individual may have a similar origin, in different responses of cells to identical prepatterns, although in this case the differences in response are not genetically determined, but arise in the course of embryonic differentiation. However, these successes in explaining variation in adult structure in terms of varying responses to unchanging prepatterns carry with them the danger that prepatterns may come to be regarded in a somewhat mystical light. It is therefore an important feature of the ocelli-less mutant that some of the abnormalities to which it gives rise can be interpreted only as the results of changes in the prepattern.

One question which it was hoped that this investigation would answer is whether the developmental mechanisms responsible for patterns which are constant in almost all members of a species have anything in common with those responsible for patterns which vary from individual to individual. The kinds of variation in the arrangements of microchaetes and of macrochaetes which have been described support the idea that the processes which determine the positions of the two types of bristle are similar. If so, the relative constancy of macrochaete patterns presumably arises because the number of macrochaetes in any particular pattern is small. It has been argued at length elsewhere[1] that mechanisms formally similar to that suggested by Turing can give rise to a constant pattern only if the number of peaks is small (approximately 5 to 7). The essence of the argument is that the number of structures formed will be the nearest integer to the ratio of the size of the field to the 'chemical wavelength'; consequently the larger the number of structures which is to be kept constant, the smaller must be the coefficient of variation of this ratio. The simplest method of ensuring the constancy of larger numbers is by a process which was called 'multiplication'. The morphogenetic field is first divided by one patterning process into a small number of large regions, and then subdivided by a second process into a larger number of smaller regions. It is therefore interesting that Ursprung[12] has shown that the development of the genital imaginal disks of *Drosophila* has a stepwise character of this kind. In more general terms, one reason for the stepwise nature of so many developmental processes may be that only processes of such a kind can give rise to uniform results.

If a number of structures are arranged in a linear series, as are, for example, a series of segments, a multiplicative process requires that two patterning processes be separated in time, occurring one after the other. But if the structures are arranged on a surface, another type of multiplicative process is possible, with an equivalent gain in accuracy. Two patterning processes can occur simultaneously, but along different axes; one process can determine the number of 'rows' and a second the number of structures in each row. This requires that the morphogenetic field should be initially anisotropic, whereas Turing supposed the field to be isotropic. It is therefore interesting that Weiss[13] has shown that the regular arrangement of fish-scales depends on a pre-existing anisotropy; that is, on the presence of two sets of collagen fibres at right angles to one another. In the cuticle of *Rhodnius*, on which the only structure visible on the adult tergites is a series of transverse ripples, Locke[14] has demonstrated the presence both of an antero-posterior gradient and of a side-to-side polarity. The arrangement of microchaetes on the sternites of *Drosophila* suggests that the field is isotropic; compare the transverse rows, figure 1*d*, with the diagonal rows, figure 1*e*. But on the dorsal surface of the thorax there are reasons for supposing that the 'rows' and 'columns' are separately determined. There is a constant number of antero-posterior rows of microchaetes, the spacing between rows being appreciably greater than between bristles within a row. A connexion between the mechanisms responsible for the arrangements of macrochaetes and of microchaetes is also indicated, since the two pairs of dorsocentral macrochaetes always occur in the fifth row of microchaetes, counting from the mid-dorsal line, and each macrochaete occupies a position in the row which would otherwise be occupied by a microchaete.

Summary

The bristles of *Drosophila* fall into two classes, the microchaetes which are small, numerous, and which vary in number from individual to individual, and the macrochaetes, which are larger, fewer in number, and constant in arrangement within a species, although many mutants which alter their number are known. Variations in the arrangement of macrochaetes in populations of *D. subobscura* homozygous for the mutant ocelliless are interpreted in terms of variations in a 'prepattern' determining the positions of the bristles, and of the competence of cells to respond to this prepattern by forming bristles. A process whereby such a prepattern may develop is described. It is argued that the arrangement of the microchaetes is determined by a similar process, differing only in that it is less accurately regulated. Mechanisms which may increase the accuracy of prepattern formation are discussed.

Acknowledgements

This work was supported by a grant from the Nuffield Foundation, whose assistance is gratefully acknowledged. Our thanks are also due to Mrs Sheila Maynard Smith, who provided the solutions of Turing's equations shown in figure 1.

References

1 J. Maynard Smith (1960) Continuous, quantized and modal variation. *Proc. roy. Soc. B*, **152**, 397–409.

2 C. Stern (1956) Genetic mechanisms in the localized initiations of differentiation. *Cold Spr.Harb. Symp. quant. Biol.*, **13**, 375–82.

3 A. M. Turing (1952) The chemical basis of morphogenesis. *Phil. Trans, roy. Soc. B*, **237**, 37–72.

4 J. Maynard Smith & K. C. Sondhi (1960) The genetics of a pattern. *Genetics*, **45**, 1039–50.

5 V. B. Wigglesworth (1959) *The Control of Growth and Form*. Ithaca, New York: Cornell University Press.

6 P. H. Sengel (1958) Recherches expérimentales sur la différenciation des germes plumaires et du pigment de la peau de l'embryon de Poulet en culture *in vitro*. *Ann. Sci. nat.*, **20**, 431–514.

7 C. H. Waddington (1956) *Principles of Embryology*. London: Allen & Unwin.

8 K. C. Sondhi (1961) Developmental barriers in a selection experiment. *Nature*, **189**, 249–50.

9 K. C. Sondhi (1961) Selection for a character with a bounded distribution of phenotypes in *Drosophila subobscura*. *J. Genet.*, **57**, 193–221.

10 K. C. Sondhi (1962) The evolution of a pattern. *Evolution*, **16**, 186–91.

11 H. Kroeger (1958) Uber Doppelbildungen in die Leibeshöhle verpflanzter Flügelimaginalscheiben. von *Ephestia kuhniella Z. Roux. Arch. EntwMech. Organ.*, **150**, 410–24.

12 H. Ursprung (1959) Fragmentierungs- und Bestrahlungsversuche zur Bestimmung vom Determinationszustand und Anlageplan der Genitalscheiben von *Drosophila melanogaster*. *Roux. Arch. EntwMech. Organ.*, **151**, 504–48.

13 P. Weiss (1959) In *Biological Organization, Cellular and Subcellular*. London: Pergamon Press.

14 M. Locke (1959) The cuticular pattern in an insect, *Rhodnius prolixus* Stål. *J. exp. Biol.*, **36**, 459–77.

Eugenics and Utopia

There is no field of application of science to human affairs more calculated to arouse our prejudices than eugenics. I cannot hope to be free from these prejudices, but in this essay I will try to separate what we ought to do from what we can now do and from what we may in the future be able to do. These problems should be thought about because our ability to alter the future course of human evolution is likely to increase dramatically during the next hundred years.

There are three ways in which we may be able to alter man's biological capacities, which I shall call selectionist eugenics, transformationist eugenics, and biological engineering. Briefly, selectionist eugenics is the application to ourselves of the techniques which, since the Neolithic revolution, we have been applying in the breeding of our domestic animals and plants. In effect, we take measures to ensure that individuals with characteristics we like will contribute more to future generations than individuals with characteristics we dislike. These measures range from the simple to the sophisticated, from the slaughter of runts to the cold storage of spermatozoa for artificial insemination. The development of a science of population genetics has enabled us to estimate with greater accuracy the consequences of any particular interference with the breeding system and to choose between effective and ineffective methods of selection. But it has not altered the fact that this is an extremely slow and inefficient method of altering the genetic properties of a population and one whose speed can be increased only by increasing the intensity of selection; a bigger change is produced in the properties of the next generation if ninety-nine per cent of the males in the present one are selectively slaughtered or sterilized than if one per cent are so treated.

Recent advances in molecular genetics have raised the possibility of a different and far more effective method of genetic change which I shall call transformationist eugenics.

At present, if we wish to eliminate an undesirable gene from a population, our only method of doing so is to reduce the breeding chances of those individuals carrying the gene; but now that we know something of the chemistry of heredity, it is possible to think of the direct alteration or transformation of particular genes. Today, this can be done only in micro-organisms, and then in only a very small proportion of the cells exposed to the transforming agent. But it will be surprising if direct gene transformation does not become possible in man and higher animals during the next hundred years. If so, it will increase by many orders of magnitude the speed and economy with which the genetic properties of populations can be changed.

Finally, the continued development of surgical techniques, together with chemical methods of altering development and with tissue and organ culture, will make it possible to produce quite profound alterations in the biological properties of individuals without altering their genetic constitution. This I shall call biological engineering.

Before discussing in greater detail the technical and ethical considerations raised by these methods, there is one general point to be made: man is evolving anyway. That is to say, changes are taking place in the genetic properties of the human population whether we like it or not; and almost every political and social measure we take influences to some degree the nature and direction of these changes.[1] Probably the most important changes at the present time are due not to selective survival or fertility but to changes in the breeding structure consequent on increased population size and increased mobility. At one extreme is the reduction in the frequency of marriages between close relatives, which is likely at least in the short term (that is, for some hundreds of years) to have beneficial effects by reducing the frequency of diseases caused by recessive genes; at the other extreme is the increase in the frequency of inter-racial marriage, although social pressures have both minimized this change and rendered it almost impossible to evaluate its consequences accurately.

But in the literature of eugenics, more attention has been paid to selective effects, and in particular to the differential fertility of social classes and the consequences of improved medical care. The importance of the former subject has probably been exaggerated—the observed differences may not last long enough to have significant evolutionary consequences —but it is comforting in any case that measurements of IQ in 1932 and 1947 in Scotland showed a slight but significant rise on the second occasion although the negative correlation between the IQ score of children and the size of the family to which they belonged had led to the prediction that the mean IQ of the population must decline. Even if, as seems likely, the observed rise was the result of particular environmental rather than genetic factors, at least no measurable genetic decline has occurred.[2]

The effects of improved medical services should perhaps be taken more seriously because these are likely to be long-lasting. The argument is as follows. Improved medical and social care make it possible for people who in the past would have died to survive and have children. In so far as their defects were genetically determined, they are likely to be handed on to their children. Consequently, the frequency of genetically determined defects in the population is likely to increase; and an increasingly large proportion of the population will be engaged in keeping the rest alive. I think we have to accept the fact that there is some truth in this argument, but it is a little difficult to see what we should do about it. To ban the manufacture of glasses and the administration of insulin because these activities permit astigmatics and diabetics to breed seems inhumane. If we are going to administer insulin, there is no rational gound for refusing to undertake the more expensive and time-consuming job of feeding babies who suffer from a genetically determined form of mental defect known as phenylketonuria on a diet free of phenylalanine. And as our knowledge increases, the number of tasks of this kind which we shall feel obliged to undertake will increase. An extreme eugenist might suggest that although we should not ban the administration of insulin we should insist on the sterilization of those who require it. The difficulty with such a policy is that almost every human being possesses at least some characteristics, physical, moral, or intellectual, which we would prefer not to be transmitted to the next generation. The only humane answer at present appears to be that an increasing investment in medical and social care is a price we should be prepared to pay.

But there are two mitigating circumstances. First, as I shall discuss later, there may be a long-term way out of this dilemma.

Second, not all the genetic consequences of improved medical services are dysgenic, and it may even be that most are not. To see why this is so, consider the case of malaria and sickle cell anaemia. Briefly, there exists in man a gene S, which in *homozygous* condition causes fatal anaemia. (An individual who inherits the same gene, defective or otherwise, from both parents is said to be homozygous; an individual inheriting different genes from his two parents is said to be heterozygous.) Yet this gene occurs in high frequency in certain places, particularly in parts of Africa where malaria is or was until recently a common cause of death. The reason for this distribution is that individuals heterozygous for the gene S are resistant to malaria. The frequency of S in Negroes of West African origin now resident in America is lower than those in West Africa, presumably because in the absence of malaria natural selection against homozygotes has not been balanced by selection in favour of heterozygotes; a similar decrease in frequency of S is likely to occur in Africa as malaria is brought under control. This is an evolutionary change consequent on improved medical care (that is, the eradication of malaria), but it is a desirable change since it has reduced the number of children dying of anaemia.

The case of sickle cell anaemia and malaria is rather a special one, but beneficial evolutionary consequences of improved medical care may be quite common. The greatest change so far produced by medicine (more particularly by preventive medicine) on the pattern of mortality is the reduction in the number of people dying of infectious disease. This has led to a reduction in the selection pressure in favour of disease-resistance and presumably to an increase in susceptibility. As long as we continue to control infectious disease by improved hygiene, inoculation, and so forth, this is probably a good thing. The reason for this is that in evolution, as in other fields, one seldom gets something for nothing. Genes which confer disease resistance are likely to have harmful effects in other ways; this is certainly true of the gene for sickle cell anaemia and may be a general rule. If so, absence of selection in favour of disease resistance may be eugenic.

Now that death from infectious disease is rare in industrial countries, the main efforts in medical research are concentrated on diseases such as cancer and rheumatism, which usually affect older people. Cures for cancer (other than leukaemia) would not have significant genetic consequences because, although they would prolong the life of many people, they would seldom add to the number of children born to such people. The main dysgenic effects of medical progress arise from cures of defects which are present at birth or appear before reproductive age, and in whose causation there is a genetic component.

So far, I have considered genetic changes which are happening or are likely to happen as unintended by-products of measures undertaken with other ends in view. I now turn to the possible effects of intentional eugenic measures, first considering measures of selective eugenics which either are already technically possible or are fairly certain to become possible in the near future. Such eugenic measures can be 'negative', that is, concerned with the elimination or reduction in frequency of undesirable traits, or 'positive', that is concerned with improving the average performance, or the proportion of individuals capable of outstanding performance, of socially desirable tasks.

The probable effectiveness of negative eugenic measures depends first on whether the characteristic in question is genetically determined and, if so, whether it is caused by a *dominant* or a *recessive* gene. (A dominant gene is one which produces an effect in single dose, being inherited from one parent only; a recessive gene has no effect unless inherited from both parents, or at least has no serious effect.) For characters caused by dominant genes, negative eugenic measures could be effective but are usually pointless; for characters caused by recessive genes, they are ineffective.[3]

Consider first a character, Achondroplasia (dwarfism with short legs and arms), which is caused by a dominant gene. Since the gene is dominant, we can recognize all carriers of the gene. If we so wished, we could by sterilizing all achondroplastic dwarfs prevent any carriers from passing the gene on to the next generation. This would not completely eliminate the character from the population but would reduce its frequency to the number of cases arising by new mutation in each generation. But why should we wish to do this, since there is little to prevent an achondroplastic dwarf from leading a contented and useful life? If a dominant mutation is lethal or seriously disabling, selection will keep it at a frequency close to the mutation rate without our intervention; if it is not disabling, why should we interfere? Unhappily, there are exceptions to this easy excuse for inactivity. An example is Huntingdon's chorea, which is due to a dominant gene which does not manifest itself until middle life, after the affected person has had children. The condition differs from Achondroplasia in being fatal and very distressing for the sufferer. Additional distress arises if relatives of affected persons foresee, correctly, that they may develop the disease. In such a case, there are good grounds for discouraging any person with even one parent or sibling who has developed the disease from having children. I am satisfied that such people should be encouraged to undergo sterilization but doubt that sterilization should be compulsory; the case for compulsory sterilization will be stronger when we learn to recognize heterozygotes before the disease develops.[4]

Now consider a disease such as phenylketonuria, which is due to a gene which is recessive in the sense that only homozygotes suffer from the associated mental defect.[5] Approximately one person in 60,000 in Britain is homozygous for the gene; it follows that one person in 122 is a carrier.[6] Consequently, if we could not recognize the heterozygotes, then the sterilization of homozygotes (in fact, untreated homozygotes normally do not get married or have children) would remove only 1/245 of the mutant genes per generation. Eugenic measures would therefore be ineffective unless applied to heterozygotes, who can in this case be recognized biochemically although they are not mentally defective. But it seems likely that most people are carriers of at least one lethal or deleterious gene, although they cannot at present be recognized as such. It follows that as our ability to recognize heterozygotes increases, we could be led to sterilize almost the whole population on eugenic grounds, which is clearly absurd.

The ability to recognize heterozygotes for such conditions as phenylketonuria makes it possible in principle to eliminate the affected homozygotes by preventing marriage between heterozygotes. (The statement in the previous paragraph that almost everyone is heterozygous for something does not invalidate this conclusion: all that has to be avoided is marriage between two people heterozygous for the *same* gene; and this would rule out only a small fraction of possible marriages.) It is admittedly difficult to see how this can be achieved, but a start might be made by testing relatives of affected persons and partners in prospective marriages between cousins. There is also a sense in which such a measure would be dysgenic. By preventing the birth of affected individuals, it would remove any selection against the mutant gene; and this would lead to an increase in its frequency in the population.[7] To this extent, preventing marriages between heterozygotes could have dysgenic effects similar to the cure of a genetically determined disease.

In earlier discussions of eugenics, suggested measures of positive eugenics took the form of legislation designed to encourage particular classes of persons to have more children. Two examples of such suggestions are increased family allowances for university teachers and a tax on children, the logic behind the latter suggestion being that only the rich would be able to afford children and that wealth is at least an approximate measure of genetic worth.[8] But in recent years increasing attention has been paid to the possibility of artificial insemination. H. J. Muller[9] and, more recently, Julian Huxley[10] have suggested that we should try to persuade married women who have had one child by their husbands to have a second child by a donor of their choice. In view of the sources from which it emanates, if for no other reason, this suggestion merits careful examination.

First, how effective would such a measure be? I shall discuss the effects on a single metrical character; to be concrete I shall consider IQ score, since intelligence is the quality most usually prized by people in academic circles who propose eugenic measures. I shall make the following assumptions, which appear to be optimistic:

(1) Among women, one per cent could be persuaded on eugenic grounds to have half their children by artificial insemination.

(2) The husbands of such women would be a random sample of the population.

(3) The mean IQ of the donors chosen would be one standard deviation above the population average. (Without intending to be either facetious or offensive, it is fair to ask what would be the relative popularities of Francis Crick and Ringo Starr.)

(4) The realized heritability of IQ scores is 0·5. (The realized heritability is defined as the progress under selection divided by the selection differential; that is to say, in the case

of IQ score, it is the increase in the mean IQ score in a generation divided by the difference between the mean IQ of the selected parents and that of the population from which they were selected. The realized heritability normally lies between 0 and 1, and the value of 0·5 is fairly typical for a metrical character.)

Given these assumptions, the mean IQ score of the next generation (allowing thirty years per generation) would be approximately 0·04 points higher than it would otherwise be. Compared with the rise of approximately two points observed in fifteen years in Scotland—which probably resulted from such things as the spread of radio and television sets—I doubt whether such a rise would be worth the trouble.

Artificial insemination would be less effective in man than in domestic animals because a number of conditions can be satisfied in the latter case but not in the former. These conditions are:

(1) It is possible to define the objective of selection—for example, growth rate or milk yield—and to accept deterioration in other characteristics—for example, mobility or intelligence.

(2) It is possible to choose the male donors on the basis of this objective and to use progeny testing to ensure that the donors pass the appropriate characters on to their children.

(3) It is possible to ensure that most females have most of their offspring by artificial insemination.

I assumed above that only one per cent of women could be persuaded to accept artificial insemination on eugenic grounds. Clearly, the effectiveness of the procedure would be increased if the proportion of women participating were greater. There have been societies in the past in which a large proportion of the women have been persuaded or coerced into a breeding system which had genetic consequences similar to the scheme suggested by Muller and Huxley. For example, among the Nambikuave Indians of central Brazil, a chieftain, nominated by his predecessor but dependent on popular consent, is the only member of the group to have a number of wives.[11] Although this practice does not seem to have been undertaken for genetic reasons, it cannot fail to have genetic consequences.

I do not believe that a larger proportion of the world's population will ever adopt such a system, using either artificial or natural insemination, but this belief may only reveal my prejudices. But it does seem possible that a small racial or religious group might adopt such a practice. If such a group could maintain a fair degree of genetic isolation from the rest of the population and if the great majority of women in the group bore at least one child by a donor of high IQ (the argument, of course, will apply to any character), then after a century the mean IQ of the group might have risen by one standard deviation, or fifteen points. In other words, a group might arise with an average intelligence similar to that of a group of students selected for a university. This seems hardly sufficient to justify the establishment of a new religion.

But what if artifical selection were continued not for a century but for a millennium? It is unlikely that the mean IQ would rise by ten standard deviations. Experience shows that if intense artificial selection for a single character is continued for a number of generations, the genetic response, although rapid at first, tends to slow down and even to stop. It is impossible to predict at what level this 'plateau' will be reached.[12] But it seems quite likely that if a human community were to practice artificial selection for intelligence for a thousand years, there would be a rise of several standard deviations in the mean IQ and the community might contain several individuals with mental capacities greater than those of anyone alive today.

But as an estimate of what would happen if, for example, a number of groups such as the American Academy of Arts and Sciences were to campaign for artificial insemination on eugenic grounds, a rise of 0·04 points per generation seems optimistic. Nevertheless, it has been argued that artificial insemination is valuable in man because a small rise in mean score would produce a disproportionate increase in the number of people with exceptionally high scores. Thus, if it is assumed that IQ score is normally distributed and that a small change in mean IQ score does not alter the variance of the score (this need not be true, but it might very well be true),[13] then an increase of one point in mean IQ would be accompanied by an increase of twenty per cent in the proportion of people with IQs above 175. It is argued that although an increase of one point in the mean IQ might not be worth bothering about, an increase of twenty per cent in the number of geniuses is well worth striving for.

The argument is weakened by the fact that IQ score is not

normally distributed; there are many more people with very high and very low scores than would be predicted on the assumption of normality.[14] Thus, if the distribution were normal, an increase of mean IQ of one point would lead to an increase in the proportion of people with a score greater than 175 from 3·3 per million to 4 per million. But since the distribution is not normal, the actual increase would be approximately from 77 per million to 85 per million.

But the main weakness of the argument lies in the assumption that an increase in the proportion of people with IQ scores above 175 would necessarily, or even probably, be associated with an increase in the number of people of outstanding ability as judged by their achievements. If this were so, it is difficult to explain why some quite small populations, for example Periclean Athens, should in a short time have produced such a number of people who, judged by their achievements, were of outstanding ability, whereas other larger populations, such as Greece during the Byzantine empire, should have produced hardly any. This is not to imply that outstanding achievements do not require unusual genetic endowments or that anyone could have written the *Principia* if he had had Newton's opportunities. What is suggested by a comparison of Greece in Classical and in Byzantine times is that any reasonably large population is likely to contain people genetically capable of outstanding achievements if social conditions are favourable. The same point is made perhaps more convincingly by referring to the frequent occurrence of simultaneous yet independent discoveries in science. It follows that a small increase in the proportion of people with IQ scores above 175 is unlikely to be important.

So far, I have accepted the four assumptions listed above as reasonable approximations. But one of the assumptions—that the husbands of women accepting artificial insemination would be a random sample—is manifestly false for two reasons. First, women accepting artificial insemination on eugenic grounds would hardly be a random sample; and, since mating in man, for intellectual and moral characteristics at least, is not random, their husbands would be likely to resemble them. Second, if the husbands agreed—and the results if they did not would hardly be desirable—they would presumably be above average in humility and unselfishness. It is at least possible to argue that these qualities are more desirable socially than the qualities for which the donors would be chosen. If so, the measure, in so far as it had any effects, would be likely to be dysgenic.

This raises the major difficulty of all suggested measures of positive eugenics, the problem of deciding what we want. It is fairly easy to recognize characteristics—blindness, mental defect, lameness—which we would wish to avoid in our own and in other people's children but much more difficult to define characters we wish to encourage, particularly when it is remembered that these characters may be mutually incompatible. Most experience with artificial selection in animals leads to the conclusion that selection in favour of a particular character—for instance, milk yield in cattle or the number of bristles in Drosophila—is effective in altering the selected characters in the desired direction; but the alteration is accompanied by changes in many other characters, changes whose nature cannot be predicted in detail but which are usually undesirable in that they lower fertility or the probability of survival, or impair performance in other ways.[15] This is only a restatement of the point made above when discussing disease resistance, that you rarely get something for nothing. It is a point usually forgotten in discussions of eugenics.

Two points should be made concerning the problem of deciding what characteristics are desirable. First, it is probable that in man at least some desirable characteristics arise in genetic heterozygotes; if so, it is unreasonable to expect them to breed true. Second, it is far from clear that what we want is a genetically uniform population; indeed, societies seem much more likely to be workable if they contain individuals with a wide range of genetic capabilities.[16]

If our objective is to increase the proportion of genetically gifted people in the population, there is a method which is likely to become feasible in the fairly near future and which would be considerably more effective than artificial insemination. This is to make clonal 'copies' of successful people. It has already proved possible to remove the nucleus from a fertilized frog egg and to replace it with the nucleus from one of the cells of a developing embryo; the egg then develops into a frog having the genetic characteristics of the embryo from which the nucleus was taken.[17] It will perhaps soon be possible to remove a fertilized or unfertilized human egg from the oviduct, remove the nucleus, and replace it with a nucleus from, let us say, a germ-line cell of some individual whose genotype we would like to reproduce. Implanted in a uterus, this egg would then develop the same genetically determined characteristics as those of the individual from which the nucleus was taken.

Leaving aside for the moment the desirability of such a 'cloning' technique, let us turn to why it would be more effective than artificial insemination. In artificial insemination, only half the genes of the donor are transmitted. Therefore, their effects may be 'diluted out' by the genes of the mother; and if the peculiar and desired characteristics of the donor depended on interactions between genes, these are likely to be lost. But in the cloning technique, an exact genetic replica, as in monovular twins, would be obtained.

How strong are the arguments for adopting this measure, supposing that it does become practicable? I do not want to be dogmatic on this point, but two arguments against it should be mentioned. First, the arguments outlined above for believing that human populations have an adequate supply of talented people to meet the problems of the time would, if accepted, show that there is little to be gained by adopting the cloning technique. Second, people 'conceived' in this way could have severe and perhaps crippling psychological difficulties. Sons of famous fathers not infrequently suffer because too much is expected of them; much more might be expected of children known to be genetically identical to a famous 'ancestor'.

I now turn from selectionist to transformationist eugenics, from what we can do to what we may be able to do in the future. I want again to consider the case of sickle cell anaemia, although there is a risk that this condition may come to play the same distorting role in evolutionary speculation today that the neck of the giraffe did in the last century. It is known that a person homozygous for the gene S differs from normal people because the haemoglobin in their red blood cells is insoluble and that this difference is due to the substitution of the amino acid valine for glutamic acid at a particular position in the β chain of their haemoglobin. It is reasonably certain that this abnormality is due to the presence of a single abnormal base in a DNA molecule in the chromosomes of blood-forming cells and that this, in turn, is due to the presence of a single abnormal base in a DNA molecule in the fertilized egg from which they developed (strictly speaking, there must have been four abnormal bases, since there were two homologous sets of chromosomes in the egg, each containing an abnormal base pair). When the details of the genetic code have been discovered, which is likely to be soon, it may be possible to specify which base has been substituted for which—for example, that adenine has replaced cytosine at a particular place.

People heterozygous for the gene S can be recognized, since their red blood cells contain about forty per cent of the insoluble haemoglobin and about sixty per cent of normal. A baby suffering from sickle cell anaemia will be born only if two heterozygotes marry (except for new mutation or illegitimacy). As was pointed out earlier, the birth of anaemic babies could be avoided by preventing the marriage of heterozygotes. It could also be prevented if it proved possible to transform a single base—say adenine to cytosine—in the sperm cells of the father, or in the oocytes of the mother, or in the fertilized egg. This would be an example of negative transformationist eugenics. It would have the immediate effect of preventing the birth of defective children without making it necessary to interfere with the choice of marriage partners and without having the dysgenic effect of causing a gradual increase in the frequency of deleterious genes.

Of the various methods of eugenics which have been or will be discussed in the essay, there seems little doubt that negative transformationist eugenics would be the most desirable. It would require the minimum interference with who marries or has children by whom; its effects would be confined to the limited and generally acceptable objective of preventing the birth of children with specific defects; and, far from having dysgenic effects, transformationist eugenics could provide a means of counteracting the long-term dysgenic effects of some types of medical care.

The drawback, of course, is that such methods are not at present practicable; and it is not yet possible to see how such transformation could be achieved. The major difficulty is the restricted nature of the transformation required. Thus a chemical procedure which transformed all or many of the adenine molecules in a nucleus into cytosine would certainly be fatal; instead, only one particular adenine molecule among the hundreds of thousands present must be transformed. Because of two properties of nucleic acid, namely, homologous pairing and recombination, the problem is not quite so hopeless as it sounds. The first of these two properties makes the following situation possible: if a normal 'gene' (DNA molecule) for haemoglobin could be introduced into a cell carrying the mutant S gene, this normal molecule might pair base by base with the abnormal one. The second property raises the possibility that in some circumstances the normal molecule might replace the abnormal one in the chromosome. Something of this kind does occur in the phenomenon of bacterial transformation; unfortunately, it is now confined to bacterial cells, which are much readier to accept nucleic acid molecules than are animal cells, and even then it is possible to transform only a small proportion of the cells exposed to transforming nucleic acid. I find it impossible to say how much my conviction that transformation will become a practicable eugenic tool arises because the wish is father to the thought; but at least it seems rational for the next hundred years or so to attempt to cure or to make life possible for people with congenital diseases without worrying too much about the ultimate dysgenic effects.

Transformationist eugenics has its most obvious area of application in the negative field, in altering genes which give rise to obvious and gross deficiencies. It is possible to visualize positive application in animal breeding; if, for example, resistance to a particular disease, or ability to digest a particular food, could be shown to depend on the presence of a particular enzyme, then a gene determining that enzyme might be incorporated into the genotype of a domestic species. But the major application of transformationist genetics is likely to be in producing genetically changed micro-organisms designed to play particular roles in the manufacture of food and of other complicated chemical substances. It is more difficult to see positive applications to man. The production of individuals of outstanding intelligence will again be taken as an example, although even greater difficulties would arise if the characters chosen were, for instance, artistic ability or moral worth. The difficulty is simply stated: we do not know what changes in the egg's ability to synthesize specific proteins would lead to increased intelligence in the adult developing from the egg; therefore, even if we knew how to bring about specific gene transformations, we would not know what transformations to make. There is no reason to think that the problem is insoluble, but it would appear to be much further from solution than the problem of genetic transformation itself.

This brings me to the third technique available for the alteration of man's nature, that of biological engineering.[18] Here I have in mind the extension of existing medical techniques from the negative to the positive field. Today it is standard practice to attempt to cure many congenital defects by surgical or medical techniques, and there is no reason to doubt that treatment of congenital disease will be supplemented or replaced in the future by methods of treating the foetus so that the developmental process is altered and a normal child is born. But at this moment we do not use or contemplate using such techniques to produce outstanding individuals. For example, it would perhaps be technically possible through surgery to produce a man whose legs were so lengthened that he could run a mile in $3\frac{1}{2}$ minutes. But, sensibly enough, we prefer to let nature take its course and manufacture motor cars and airplanes if we want to move fast. But we do not hesitate to cure a lame child if we can.

It seems, then, that our present practice depends on a concept of normality, however difficult that may be to define. Since we are concerned here with a problem of what we ought to do rather than what is technically feasible, it is perhaps best to regard a characteristic as abnormal if it leads to a loss of function sufficient to cause its possessor to be unhappy. But we should ask also if there are circumstances in which we might wish to produce outstanding individuals. In the field of physical performance this seems unlikely, since it will always be easier to build a machine. There is, however, one exception: we cannot build machines to make us live longer. It is not at present possible to say whether we shall ever be able to produce a large increase in human life expectancy, even though we can already ensure that a larger proportion of people survive to old age. We do not at present know whether senescence is caused by a number of physiologically independent processes—in which case, even if we prevented one of these processes people would still die at much the same age of another—or whether there is one fundamental process of which the various superficial signs of senescence are merely symptoms. If the former assumption is correct, and the evidence suggests to me that it is, then a significant extension of the human life span is likely to prove very difficult.[19] It would also contribute disastrously to the present increase in world population. But should the world population problem prove soluble without war or famine, then an increase in human life span, if it could be associated with an appropriate decrease in human fertility, seems to me very desirable.

Olaf Stapledon, in his book *Last and First Men*, imagined the use of biological engineering to produce super-intelligences. Human neural tissue was permitted to grow and ramify through the corridors of a building and was supplied with sensory information and a motor output. In *Sirius* the same author imagined a dog whose intelligence, by surgical and other means, had been made equal to that of a man. These feats are not at present technically possible, but there is no reason why it should not eventually be possible to bring about a dramatic increase in the size of certain parts of a human or animal brain by influencing development. It is, of course, by no means certain that such a simple procedure would lead to an equivalent increase in intelligence; it might equally well lead to idiocy. But there is one reason to suspect that an appropriate increase in size, together with other comparatively minor changes in structure, might lead to a large increase in intelligence. The evolution of modern man from non-tool-making ancestors has presumably been associated with and dependent on a large increase in intelligence, but has been completed in what is on an evolutionary scale a rather short time—at most a few million years. This suggests that the transformation in the brain which provided the required increase in intelligence may have been growth in size with relatively little increase in structural complexity—there was insufficient time for natural selection to do more. Of course, this process may have reached its limit, and further increase in intelligence may require a major reorganization of structure, which would be difficult to bring about by 'engineering' methods.

On balance, it seems quite likely that within a hundred years or so it will be technically feasible to do the kinds of things imagined in Stapledon's books. But even if it is, it is not clear what the consequences would be. To ask oneself the consequence of building such an intelligence is a little like asking an Australopithecine what kind of questions Newton would ask himself and what answers he would give. One way of putting the problem is this: What questions could be asked or answered by a 'super-intelligence' composed of neurons which could not be asked and answered by teams of investigators given time and the assistance of computers? It is quite possible that the answer to this question is 'none'. But I suspect that if our species survives, someone will try it and see.

The subjects discussed in this essay are diverse, so I will attempt to summarize my argument and draw some general conclusions. First, evolutionary changes are constantly occurring in the human species, and most legislative or social measures we take inevitably influence the nature of these changes. Some, but by no means all, of the genetic changes consequent upon improved medical and social services are dysgenic. At present, there is little that we can do to prevent these dysgenic effects, and the proper course for us to adopt is to regard them as part of the price we pay for being civilized. In any case these genetic changes are extremely slow in comparison with technical changes, and it is reasonable to hope that before they have become significant it may be possible to avert or reverse them by techniques of genetic transformation.

Deliberate measures to alter man's biological nature may be negative, designed to prevent or cure mental or physical defect, or positive, designed to produce individuals of unusually high performance in a desired area or to raise the mean level of performance. Techniques available can be classified as selectionist eugenics, transformationist eugenics, and biological engineering. Selectionist eugenics involves altering the relative number of offspring born to particular kinds of individuals or pairs. In most cases, these procedures are likely to be too ineffective to be worth bothering with. But it is worth making an effort to prevent individuals who carry deleterious dominant mutations which manifest themselves late in life from having children and to prevent the carriers of the same recessive lethal or deleterious gene from marrying one another, although the latter measure would have dysgenic effects in the long run. In the positive field, selectionist eugenics is again likely to be relatively ineffective. Probably the most effective procedure, and one which should become technically feasible in the fairly near future, would be some form of cloning.

Transformationist eugenics, involving direct alteration of specific genes in specific ways, is not at the moment possible, but may become so. If it does become possible, its use in negative eugenics would be desirable; but it is less clear what role it could play in positive eugenics.

Biological engineering in the negative field is simply another word for current medical practice. Problems, both technical and ethical, arise in the use of similar techniques to produce individuals of outstanding ability rather than to cure or prevent abnormality. Two major undertakings can be considered. One, a significant increase in the human life span, although dangerous unless the world population problem has been solved, will in the long run be desirable, but it is likely to prove very difficult and perhaps impossible. The other, the production of individuals generally resembling human beings but of outstanding intelligence, may prove relatively easy, although there is no guarantee that this is so; but even if it is technically feasible, it does not seem possible to predict what important results, if any, would ensue.

But these problems of transformationist eugenics, increase of longevity, and super-intelligence still lie in a future which is distant in historical terms even if it is immediate on an evolutionary time scale. Our immediate problem is what should be done with the means now available to us, and, more immediate still, what should geneticists and other biologists recommend be done.

I think the answer to this question is that we should not recommend that anything be done except the simple and limited negative measures suggested above. The reason for this is that I believe recommendations of positive eugenic measures can at the present only distract attention from more urgent and important questions. The most urgent message which biologists have to convey to the public is that if something is not done to arrest the present increase in world population, then that increase will be arrested by war, disease, and starvation. Eugenics can wait, birth control cannot.

Notes and References

1 For a discussion of this point, *see* P.B. Medawar (1959) *The Future of Man*. London: Methuen.

2 The evidence for a negative correlation between family size and IQ is summarized by L.S. Penrose (1955) Evidence of heterosis in man. *Proc. roy. Soc. B*, 144, 203. Penrose puts forward a genetic hypothesis which would account for this correlation and yet predict no change in IQ with time. His views have been criticized by P.B. Medawar, *op. cit.*, and by K. Mather (1963)

Genetical demography. *Proc. roy. Soc. B*, 159, 106. For a comparison of the 1932 and 1947 surveys in Scotland, *see* G.H.Thomson (1949) *The Trend of Scottish Intelligence*. London: University of London Press.

3 The argument which follows is presented in J. B. S. Haldane (1938) *Heredity and Politics*. London: Allen and Unwin.

4 Another genetically determined abnormality which, although not caused by a dominant gene, could be reduced in frequency by negative eugenic measures is translocation mongolism. Individuals of either sex heterozygous for a translocation involving chromosome 21 are themselves normal, but one third of their children will be mongolian idiots and one third will be 'carriers'; only one third will be normal and likely to have normal children. Such people could be recognized if the chromosomes in a skin or blood sample were examined, and most of them would be found if all relatives of known mongols were examined. But the arguments for sterilization are perhaps less strong than in the case of Huntingdon's chorea because mongolian idiots are commonly quite cheerful and contented. In any case, sterilization would not prevent the more common form of mongolism which is due to nondisjunction in the mother.

5 It has recently been found that a homozygous baby, if recognized at birth and subsequently kept on a diet free of phenylalanine, can develop normal intelligence. This does not affect the argument concerning the ineffectiveness of sterilization in the case of diseases caused by recessive genes, but it does illustrate the important point that genetically determined diseases may be curable.

6 This frequency omits cases known to have consanguineous parents. The frequency of carriers has been worked out from the Hardy-Weinberg ratio, assuming random mating.

7 It could be argued that this would not be dysgenic provided that marriage between heterozygotes was prevented. However, it is unlikely that we should succeed in preventing all such marriages. Also, the proportion of marriages contra-indicated on genetic grounds would increase. But it seems likely that long

before these effects become serious some technique of negative transformationist eugenics will be available.

8 The idea that financial measures might be used for eugenic purposes was put forward by R. A. Fisher (1930) *The Genetical Theory of Natural Selection.* London: Oxford University Press. The suggestion of a tax on children was made, perhaps not very seriously, by F. H. C. Crick (1963) in *Man and his Future,* p. 276 (ed. Gordon Wolstenholme). London: Churchill. The suggestion has the virtue of bringing out the necessary contradiction between financial measures suggested on eugenic grounds and those suggested by the humanitarian desire to protect children from the incompetence of their parents. On the same occasion, Crick made the more important point that the time has come to question our present assumption that people have a right to have children.

9 H. J. Muller (1935) *Out of the Night.* New York: Vanguard Press; *and* Genetical progress by voluntarily conducted germinal choice in Gordon Wolstenholme (ed.), *op. cit.*

10 J. S. Huxley (1962) *Eugenics in Evolutionary Perspective.* London: Eugenics Society.

11 C. Levi-Strauss (1961) *A World on the Wane.* London: Hutchinson.

12 For the occurrence of 'plateaus' in selection experiments, *see* for example, K. Mather and B. J. Harrison (1949) The manifold effects of selection. *Heredity,* 3, 131; and I. M. Lerner (1954) *Genetic Homeostasis.* New York: Wiley.

13 A small change in mean will not alter the variance if the effects of different genes on IQ are additive and if the frequencies of alleles for high and low intelligence are on the average equal.

14 C. Burt (1963) Is intelligence distributed normally? *Br. J. statist. Psychol.,* **16,** 175.

15 The occurrence of such correlated changes is not in doubt, although their explanation is still a matter of controversy; the subject is discussed in the references given in note 12 above.

16 These two points are too important to be dismissed in a brief paragraph; my excuse is that they have been dis-

cussed more fully by T. Dobzhansky (1962) *Mankind Evolving*. New Haven, Connecticut: Yale University Press.

17 T. J. King and R. Briggs (1956) Serial transplantation of embryonic nuclei. *C.S.H. Symp. quant. Biol.*, 27, 271.

18 Some possible developments of biological engineering or 'euphenics', were discussed by J. Lederberg, Biological future of man, in Gordon Wolstenholme (ed.), *op. cit.*

19 For a discussion of this point, *see* G. C. Williams (1957) Pleiotrophy, natural selection and the evolution of senescence. *Evolution*, 11, 398; and J. Maynard Smith (1962) The causes of ageing *Proc. roy. Soc. B*, 157, 115.

The Status of Neo-Darwinism

By Darwinism is meant the idea that evolution is the result of natural selection. Neo-Darwinism adds to this idea a theory of heredity. In its most general form, the theory of heredity is Weismannism, that is, it is the theory that changes in the hereditary material are in some sense independent of changes in the body or 'soma'. In particular, the theory of heredity is Mendelian, that is, it assumes that heredity is atomic, and obeys either Mendel's laws or some modification of them explicable in terms of the behaviour of chromosomes (for example, linkage, polyploidy).

There are two reasons for discussing neo-Darwinism at this conference. The first is that only in the study of evolution is there a body of biological theory in any way comparable to the theories of physics; a conference on theoretical biology can hardly refrain from discussing it. The second is that the theory in at least some formulations is tautological. 'The survival of the fittest' appears to mean merely that survivors survive. There seems little point in trying to explain evolution by a tautology.

In this article, therefore, I shall attempt first to formulate the theory in a non-tautological form. I shall then discuss what types of observation might refute it, because this is the best way of seeing whether the attempt at a non-tautological formulation has been a success. Finally, at a less philosophical level, I shall discuss what problems the theory can cope with, what problems are at present unsolved because of a lack of data or of adequate mathematical tools, and what problems seem at present inaccessible to solution without introducing new concepts.

The formulation of neo-Darwinism

The main task of any theory of evolution is to explain adaptive complexity, that is, to explain the same set of facts which Paley used as evidence of a Creator. Thus if we look at an organism, we find that it is composed of organs which are at the same time of great complexity and of a kind which ensures the survival and/or reproduction of their possessor. Evolution theory must explain the origin of such adaptations.

At the outset we are faced with a difficulty; we have no way of measuring the degree of complexity of a structure. Thus although most of us would readily agree that the organs of a man are more complex than those of an amoeba, and those of an amoeba more complex than those of a bacterium, we have no agreed criteria on which to base this decision, and no way of deciding by how much one organism is more complex than another.

It may therefore seem odd to start formulating a theory of evolution by introducing a term which cannot be fully defined. However, I see no escape from doing so. If organisms were not both complicated and adapted, living matter would not differ from dead matter, and evolution theory would have nothing to explain.

Evolution is explained in terms of three properties, multiplication, heredity, and variation, which organisms can be observed to possess. They will be considered in turn.

(1) *Multiplication.* All living organisms are capable of increasing in numbers in at least some environment. Multiplication is necessary because without it natural selection is impossible; you cannot cull a herd which can only just maintain its numbers. It is a corollary of this condition that life must consist of individuals and not of a continuum if it is to evolve.

(2) *Heredity.* Briefly, like must beget like. More precisely, before we can say that entities have heredity, a number of different kinds of entities, A, B, C, and so on, must exist, and each must tend to produce offspring like itself. Thus fire, if supplied with fuel, will multiply; but it does not have heredity, because the nature of a fire is determined by the fuel it is burning, and not by the nature of the fire from which it was lit.

(3) *Variation.* If heredity were perfect, evolution would be impossible. Occasionally an offspring must differ from its parent. Viewed in this light, variation is merely the unreliability of heredity, and as Pattee[1] has emphasized, the problem is to explain why the reliability of replication is so high, not why mistakes are sometimes made.

However, this is not the whole story. If variation is to lead to evolution, then some variations must alter 'fitness', and at least some of these must increase fitness. By fitness is simply meant the probability of survival and reproduction. A melanic moth is, by definition, fitter if it is more likely to survive, and a myopic man may be fitter if his myopia enables him to escape the draft. [Much confusion has arisen because 'fit' is not used in this sense in the phrase 'the survival of the fittest'. If it were so used, the phrase would indeed be tautological. A more precise though less elegant (and hence less 'fit') phrase would be 'the survival of the adaptively complex', that is, organisms are adaptively complex or, as Bohm might say, 'harmonious', because such organisms survive better than less harmonious ones.] It follows from this definition that fitnesses can be compared only in a specified environment or range of environments.

Given entities with these properties, variants of higher fitness will replace their less fit ancestors: according to neo-Darwinism, this replacement constitutes evolution. Very early in evolution there arose a distinction between 'genotype' and 'phenotype', because those genotypes which gave rise to a phenotype were fitter than those which did not. By 'genotype' I mean that part of an organism which is replicated; by 'phenotype' I mean a structure or sequence of structures developing under the instructions of the genotype, and whose function it is to ensure the replication of the genotype. (These are not quite the accepted meanings of the words in genetics, but I would rather misuse words than invent new ones.) To paraphrase Butler's unexpectedly perceptive remark that the chicken is the egg's means of ensuring the production of another egg, the phenotype is the genotype's way of ensuring the production of another genotype. Once there is a distinction between phenotype and genotype, there is a process of epigenesis, and a process of decoding whereby the instructions in the genotype are translated into the structure of the phenotype.

Something must now be said about the origin of new variations, that is, of mutation. It has been said that mutation is 'random'. Apart from the difficulty of defining the word, the statement is in one sense untrue, because different mutagenic agents produce different kinds of change in the genetic material.

Observation suggests that two things are in fact true about mutation: (1) most mutations lower fitness. If this were not so, evolution would proceed without natural selection; (2) if a variant phenotype arises because development occurs in a changed environment, this will not produce corresponding changes in the genotype, such as to give rise in the next generation to the variant phenotype. This is the Weismannist assumption, expressed colloquially by saying that acquired characters are not inherited. Note that it does *not* say that changes in the phenotype cannot cause mutations, because of course they can. The apparent randomness of mutation arises because genotype and phenotype are connected by an arbitrary code.

The Weismannist assumption is expressed in molecular terms in the 'central dogma', which states that information can pass from DNA to protein, but not from protein to DNA; more precisely, if a new kind of protein is introduced into a cell, this cannot direct the synthesis of a new DNA molecule able to direct the synthesis of more of the new protein.

So far I have been describing a set of properties of organisms or, more precisely, a set of properties which neo-Darwinism assumes all organisms to have. This is not by itself a theory of evolution. The theory of neo-Darwinism states that these properties are necessary and sufficient to account for the evolution of life on this planet to date.

The limitations of time and place are important: of time, because in future we shall doubtless control our own evolution and that of our domestic animals and plants by the direct biochemical manipulation of DNA; of place, because we have as yet no grounds for asserting that if evolution has occurred elsewhere in the universe, it has done so by neo-Darwinist processes, although I would be willing to conjecture that it has.

It may help to clarify my position if I say that I accept Bohm's argument[2] that to understand biological function we must appeal to what is beyond function, and also his statement that 'current metaphysics' (=neo-Darwinism) appeals to survival value as the trans-functional feature. Where I think he goes wrong is in regarding this procedure as tautological. He has been misled by the phrase 'the survival of the fittest'. Of course Darwinism contains tautological features: any scientific theory containing two lines of algebra does so. That it is not tautological *in toto* is best demonstrated by showing that it can be falsified, as I will now try to do.

If this formulation of neo-Darwinism is not tautological, it must be possible to suggest observations which would refute it. Such observations could take two forms: (1) it could be shown that the assumptions made by neo-Darwinism are not in fact true of all organisms; (2) patterns of evolution may occur which are inexplicable on the neo-Darwinist assumptions.

The possibilities will be considered in turn. It seems unlikely that we can show that organisms do not multiply or do not vary. However, the assumptions about heredity and about the origin of new variation could readily be disproved if they are false. Thus it should be possible to demonstrate Lamarckist effects if they occur, or 'inertial' effects whereby if one mutation in a given direction has occurred the next mutation is more likely to be in the same direction, or even 'teleological' effects whereby a succession of mutations occur which are individually non-adaptive but which together adapt the organism to a new environment.

By and large, such types of mutational events seem not to happen, and it is difficult to see in molecular terms how they could happen. It is impossible to *prove* that they do not, just as it is impossible to prove that heat never flows from a cold body to a hot one. All one can do is to assume they don't until someone demonstrates that they do.

I will now turn to the possibility that patterns of evolution occur which cannot be explained by neo-Darwinism. The first possibility is that evolutionary changes occur more rapidly than can be explained by neo-Darwinism. This would be quite easy to demonstrate if it occurred on a small scale in the laboratory. Thus suppose, for example, a population of fruit flies were kept at an unusually high temperature. By measuring the genetic variance of temperature tolerance in the population before starting, it would be possible to predict the maximum rate at which temperature tolerance would increase. If in fact it increased faster than this, then the population would have evolved by a mechanism other than neo-Darwinism.

However, most critics of neo-Darwinism accept that the theory works at the level of laboratory experimentation. They suggest instead that there are large-scale features of evolution which call for additional types of explanation. Here we are up against the difficulty that we do not understand epigenesis, and we therefore do not know how many mutations would be necessary, for example, in the genotype of a small dinosaur to turn it into a bird. Therefore we do not know how many generations of selection would be needed to produce the change. This difficulty, combined with the imperfections of the fossil record, mean that we are unlikely to be able to disprove neo-Darwinism by showing from an examination of the fossil record that evolution has proceeded too rapidly. All one can say is that where we do have a reasonably continuous record, the observed rates of change are many orders of magnitude slower than those which can be produced in the laboratory.

If, however, neo-Darwinism were false, one would expect to be able to demonstrate its falsity by examining the end-products, that is, existing organisms. Thus it follows from neo-Darwinism that if we find an adaptively complex organ, then the organ will contribute to the survival or reproduction of its possessor. One apparent exception arises in cases such as the worker bee, which have organs favouring the survival of their close relatives; but since their close relatives share many of their genes, this is explicable on the grounds that the phenotype of the worker bee ensures the multiplication of its own genotype.

If one invents counter-examples, they seem absurd. Thus if someone discovers a deep-sea fish with varying numbers of luminous dots on its tail, the number at any one time having the property of being always a prime number, I should regard this as rather strong evidence against neo-Darwinism. And if the dots took up in turn the exact configuration of the various heavenly constellations, I should regard it as an adequate disproof. The apparent absurdity of these examples only shows that what we know about existing organisms is consistent with neo-Darwinism. It is of course true that there are complex organs whose function is not known. But if it were not the case that most organs can readily be understood as contributing to survival or reproduction, Darwinism would never have been accepted by biologists in the first place.

Thus there are conceivable observations in the fields of genetics, of evolutionary changes in the laboratory, and of physiology, which could disprove neo-Darwinism. In palaeontology, although there is perhaps no possibility of a formal disproof because in our present state of ignorance about epigenesis it would always be possible to argue that a sudden evolutionary change was due to a single mutation, there are in practice many conceivable observations which would throw grave doubts on the theory. It therefore seems to me absurd to argue that the theory is tautological, though I readily admit that it is often formulated tautologically.

At present there are in my opinion no adequate observational grounds for abandoning the theory. This is of course no reason for not seeking for such grounds—I am all for people looking for Lamarckian effects, or for exceptions to the central dogma. But in the meanwhile, the theory explains so much that it is impossible to operate in biology without accepting it, just as it is impossible to operate in physics without accepting Newtonian mechanics, or some other theory which subsumes Newtonian mechanics as a special case.

The successes and failures of neo-Darwinism

I have suggested that neo-Darwinism has not as yet been refuted. But is it of any interest? Does it tell us anything not immediately obvious? Does it solve problems? Since the ability to solve problems seems to me one of the essential characteristics of a scientific theory, these questions are important. Perhaps the easiest way of answering them is to list some of the problems which can be thought about within the context of neo-Darwinism, and which would be unanswerable in any other context. This is not to say that all these problems have been solved—I would say that the first four have been largely solved, the fifth only partly solved, that we lack essential data for the solution of the sixth, that there are accepted but partially erroneous solutions to the seventh and eighth, and that both conceptual difficulties and lack of information prevent the solution of the last.

(i) How rapidly will gene frequencies change under selection?
(ii) How can one predict the effects of selection for a continuously varying character?
(iii) What processes are responsible for the genetic variability of sexually reproducing species?
(iv) How many selective deaths are needed to replace one gene in a population by another?
(v) Will selection bring genes affecting the same character on to the same chromosome?

(vi) Can selection be responsible for the evolution of characters favourable for the species but not for the individual?

(vii) Can one species divide into two without being separated by a barrier to migration?

(viii) In what circumstances will sexual reproduction accelerate evolutionary change?

(ix) Has there been time since the pre-Cambrian for selection to program the length of DNA known to exist in man?

These problems—except for the last—illustrate the field within which neo-Darwinism has been successful. Even when problems are unsolved, this is because of a lack of data or of mathematical technique rather than of concepts.

The failures of neo-Darwinism arise because of the absence of theories in the adjacent fields of epigenesis and of ecology. Lacking a theory of epigenesis, we cannot say how many gene substitutions are required to convert a fin into a leg, or a monkey's brain into a human one. Consequently we cannot say how many generations of selection of what intensity were needed to produce those changes. There is one exception to this statement of ignorance. Knowing the genetic code, we also know how many mutational steps are needed to convert one protein into another. Consequently we can speak with more precision about the evolution of proteins than about the evolution of legs or brains.

The difficulties which arise from our ignorance of ecology can best be illustrated by discussing the related problem of whether neo-Darwinism can explain the evolution of increasing complexity. Neo-Darwinism predicts that *in the short term* individuals will change in such a way as to increase their fitness in the environment or range of environments existing at the time. This may lead to an increase or a decrease in complexity. Sometimes, as in the evolution of tapeworms or viruses, it has led in the direction of decreasing complexity, albeit in an increasingly complex environment.

Thus there is nothing in neo-Darwinism which enables us to predict a long-term increase in complexity. All one can say is that since the first living organisms were presumably very simple, then if any large change in complexity has occurred in any evolutionary lineage, it must have been in the direction of increasing complexity; as Thomas Hood might have said,

'Nowhere to go but up'. But why should there have been any striking change in complexity? It is conceivable that the first living thing, although simple, was more complex than was strictly necessary to survive in the primitive soup, and that evolution of greater fitness meant the evolution of still simpler forms.

Intuitively, one feels that the answer to this is that life soon became differentiated into various forms, living in different ways, and that within such a complex ecosystem there would always be *some* way of life open which called for a more complex phenotype. This would be a self-perpetuating process. With the evolution of new species, further ecological niches would open up, and the complexity of the most complex species would increase. But this is intuition, not reason. It is equally easy to imagine that the first living organism promptly consumed all the available food and then became extinct.

What we need therefore is first a theory of ecological permanence, and then a theory of evolutionary ecology. The former would tell us what must be the relationships between the species composing an ecosystem if it is to be 'permanent', that is, if all species are to survive, either in a static equilibrium or a limit cycle. In such a theory, the effects of each species on its own reproduction and on that of other species would be represented by a constant or constants. We want to know what criteria these constants must satisfy if the system is to be permanent. A start on this problem has been made by Kerner[3] and Leigh.[4]

In evolutionary ecology these constants become variables, but with a relaxation time large compared to the ecological time scale. Each species would evolve so as to maximize the fitness of its members. If so, a permanent ecosystem might evolve into an impermanent one. For example, a predator-prey system might be permanent because the prey could burrow and so escape total extinction. But if the predator evolved the capacity to burrow too, the ecosystem would become impermanent.

What then are the criteria to be satisfied if an ecosystem not only is to be permanent, but is to give rise by evolution to permanent ecosystems of greater species diversity? We have no idea. But the first living organism, with its food supply, had to comprise such an ecosystem if evolution was to lead to increasing complexity.

References

1 H. H. Pattee (1968) The physical basis of coding and reliability in biological evolution. *Towards a Theoretical Biology, 1: Prolegomena*, pp. 67–93 (ed. C. H. Waddington). Edinburgh: Edinburgh University Press.

2 D. Bohm (1969) Some remarks on the notion of order, *and* further remarks on order. *Towards a Theoretical Biology, 2: Sketches*, pp. 18–60 (ed. C. H. Waddington). Edinburgh: Edinburgh University Press.

3 E. H. Kerner (1957) *Bull Math. Biophys.*, 19, 121–46.

4 E. G. Leigh (1965) *Proc. nat. Acad. Sci.*, 53, 777–83.

Time in the Evolutionary Process

This paper will discuss two topics, related to one another only in that both have to do with time and evolution. The first is whether there has been enough time for existing organisms, with their fantastic complexity, to have evolved by a process as apparently inefficient as the natural selection of chance variations. The second is whether there is any biological law which might enable us to put an arrow on time in evolutionary processes, as the second law of thermodynamics enables us to put an arrow on physical processes.

When confronted by the richness of organic life, it is a common reaction, particularly among non-biologists, to argue that natural selection is an insufficient explanation. Is it really possible that an elephant can have arisen by selection acting on random variation? It is difficult to give a confident answer to this question because we do not know how improbable an elephant is, or, what amounts to the same thing, we do not know how much genetic information is required to control the development of an elephant. We shall not be able to answer this question until we know more about the process of development. However, for the time being, it is reasonable to assume that the genetic information in the DNA of the fertilized egg is sufficient to control development. In the case of mammals, this amounts to between 10^9 and 10^{10} base pairs. There is no quantitative difficulty in seeing how such a length of DNA might have been programmed by selection since the origin of life some 2×10^9 years ago.

In recent years this argument has taken on an apparently quantitative form,[1] as follows. A typical protein is 100 amino acid residues in length. There are 20 different amino acids used in making proteins. Hence if all sequences are permitted, the total number of possible proteins of that length is 20^{100}. This is a very large number indeed, and much larger than the total number of proteins that have ever existed, or even than the total number which would have existed if the earth had been covered since the Cambrian by a layer of proteins several feet thick changing once a second.

It follows, or so it has been argued, that the chances of reaching the actual proteins in living organisms, which are beautifully adapted to their function, by a random walk is effectively zero. Natural selection does not help, because it can only ensure the survival of functional proteins if they arise; it cannot bring them into existence in the first place.

I have tried this argument out on a number of physicists during the past few years. Almost always their immediate reaction is to suggest that there must be some physical constraints on the mutation process whereby new proteins arise. Now it is almost certain that this is not the case. All the evidence suggests that any nucleotide sequence in DNA can exist and can arise by mutation; that if it does exist it will be translated by the cell into the corresponding protein; and that the protein will be made by the cell even if it is completely non-functional.

The way out of this dilemma can be best understood by analogy with a popular word game, in which it is required to pass from one word to another of the same length by changing one letter at a time, with the requirement that all the intermediate words are also meaningful in the given language. Thus WORD can be converted into GENE in the minimum number of steps, as follows:

WORD
WORE
GORE
GONE
GENE

This is an analogue of evolution, in which the words represent proteins; the letters represent amino acids; the alteration of a single letter corresponds to the simplest evolutionary step, the substitution of one amino acid for another; and the requirement of meaning to the requirement that each unit step in evolution should be from one functional protein to another. The reason for the last requirement is as follows: suppose that
The reason for the last requirement is as follows: suppose that a
protein A B C D . . . . exists, and that a protein a b C D . . . . the intermediates a B C D . . . . and A b C D . . . . are non-functional. These forms would arise by mutation, but would usually be eliminated by selection before a second mutation

could occur. Thus the double step from A B C D . . . . to a b C D . . . . would be very unlikely to occur. Such double steps with unfavourable intermediates may occasionally occur, but are probably too rare to be important in evolution.

It follows that if evolution by natural selection is to occur, functional proteins must form a continuous network, which can be traversed by unit mutational steps without passing through nonfunctional intermediates. In this respect, functional proteins resemble 4-letter words in the English language, rather than 8-letter words, since the latter form a series of small isolated 'islands' in a sea of nonsense sequences. Of course this is not to deny the existence of isolated proteins, analogous to the 4-letter words ALSO and ALTO.

It is easy to state the condition which must be satisfied if meaningful proteins are to form a network. Let x be a meaningful protein. Let N be the number of proteins which can be derived from x by a unit mutational step, and f the fraction of these which are 'meaningful', in the sense of being as good as or better than x in some environment. Then if $fN > 1$, meaningful proteins will form a network, and evolution by natural selection is possible. In estimating N it is necessary to distinguish two classes of mutations:

(1) substitutions of single amino acids, and additions or deletions of small numbers of amino acids, making only a small change to the protein; and

(2) mutations producing a major change in amino acid sequence. (Examples are frame shifts and intra-molecular inversions; for non-biologists, the relevant point is that there are mutational changes in DNA which alter simultaneously all or most of the amino acids in a protein.)

Mutations of the former type are much more likely to give rise to meaningful proteins than the latter. In the same way, a single random letter substitution in a meaningful word is more likely to give rise to a meaningful word than the simultaneous alteration of all the letters. Although frame shift mutations are known to occur, it is not known whether they have ever been incorporated in evolution. Hence it is better to take N as the number of possible substitutions of single amino acids. If all substitutions were possible in a single mutational step, N for a protein of 100 amino acids would be 1900. In practice the genetic code limits N to approximately 10^3.

Hence f must be greater than 1/1000. It does not follow that the fraction of all possible sequences which are meaningful need to be as high as 1/1000. It is probably much lower. Almost certainly, there is a higher probability that a sequence will be meaningful if it is a neighbour of an existing functional protein than if it is selected at random.

No quantitative difficulty arises in explaining the evolution of proteins if $fN > 1$. The argument of course says nothing about the origin of life, since it assumes that at least one functional protein exists as a starting point.

Before leaving this topic, it may be worth saying something about the geometry of the protein space; these ideas emerged during a conversation with Donald Glaser. We want space in which two proteins are neighbours if they can be converted into one another by a single mutation. For simplicity, I will assume that amino acid substitutions are the only possible mutations and, ignoring the code, that all substitutions are possible. I will start with the space representing all possible dipeptides. This will be represented by a 20×20 chess board, each of the 400 squares representing a different peptide. A single mutational step is equivalent to a rook's move. Any dipeptide can be converted into any other by two moves. However such a conversion may be impermissible in evolution if the intermediate is meaningless. Suppose two squares (dipeptides) are connected by a meaningful path, how long can the shortest meaningful path between them be? The answer is obviously the 38 rook moves required to travel from one corner to the opposite corner by moving one square alternately in a horizontal and vertical direction. This path would be shortened by the presence of any meaningful peptides other than those on the path.

Transferring these ideas to the space of all proteins 100 amino acids long, the space would be of 100 dimensions, and contain 20^{100} 'squares'. A mutation is still a rook's move, in any one of 100 directions. Any protein could be converted into any other by not more than 100 mutations. If two proteins are connected by a meaningful path, the shortest meaningful path between them cannot be greater than 1900 steps, and will usually be much shorter. (It was realized later that this is wrong. The maximum 'shortest meaningful path' could be longer than this by many orders of magnitude. Hence there may be large regions of the protein space as yet unexplored.)

It follows that if, since the origin of life, protein X has evolved into protein Y, it could always have done so in less than 1900 steps. If it has in fact taken more, it has taken unnecessary detours. 1900 steps require approximately 1 step per 10^6 years; most proteins have probably evolved more slowly (by factor of 10) than this.

As a convinced Darwinist, I published[2] the conclusion that $fN > 1$ before there was any direct evidence, since if it were not so evolution would not have happened. Since that time, it has turned out that many single amino acid substitutions can be made in proteins without seriously impairing their catalytic function.[3]

The second problem I want to discuss is whether there is any law which plays the same role in biology as the second law of thermodynamics plays in physics. On a short time scale, measured in days, biological processes have an obvious direction. The cell cycle typically ends in the division of a single cell into two, and only rarely, in the sexual process, in the fusion of two cells to form one. This is a necessary feature of life. Life is most conveniently defined as consisting of entities with the properties which enable them to evolve by natural selection; that is, the properties of multiplication, variation, and heredity. The reason for choosing this definition is that the apparently purposive or adaptive features which characterize living as opposed to dead matter can evolve in entities with these properties but not in their absence.

If this definition is accepted, then it is the property of multiplication which enables us to put a time arrow on biological processes. Heredity and variation are reversible; parents resemble (or differ from) their children as closely as children resemble their parents.

It is more difficult to say whether evolution as a whole has a direction. Thus suppose we are able to make observations on the members of a species at two points in time separated by millions of years, is there any way in which we could decide which set of observations was the earlier? At first sight it seems that we can do so by using Fisher's 'fundamental theorem of natural selection',[4] which states that for any population 'the rate of increase of fitness of an organism is equal to the genetic variance of fitness'. Since the variance cannot be negative, the law appears to state that the fitness of a population of organism must always increase. Thus, just as we can tell which of two states X and Y of a closed physical system is the later in time, by asking which has the greater entropy, so we should be able to tell which of two states X′ and Y′ of a population is the later in time by asking which has the greater fitness. Unhappily we cannot do anything of the kind.

The difficulty lies in the definition of the 'fitness of an organism'. The essential points can be understood by considering a parthenogenetic population consisting of two genetically distinct types, A and B, in proportion $pA:qB$. Suppose that we count As at birth, and the number of offspring, also counted at birth, produced by these As. Then the average number of offspring produced per A is the fitness w_A of A. The fitness w_B of B is similarly defined. Then the fitness w of the population is defined as $w = pw_A + qw_B$. It is this fitness w which, according to Fisher's theorem, necessarily increases.

There are three reasons why we cannot use this theorem to tell us which of two populations is the later in time.

(1) The fitness w_A and w_B, and hence w, can be defined for a particular environment only. For example, a population whose life span is short compared to a year may evolve in one direction in summer and the other in winter. Even if the physical features of the environment remain constant, the biotic features will not. For example, A may be rare but better at escaping a predator than B. If so $w_A > w_B$, and A will increase in frequency. The predator may then evolve, or change in habits, so that it is better at catching A than B. Then $w_B > w_A$, and evolution will reverse its direction. Such reversals may have been a common feature of evolution of defence against predators and disease.

(2) The 'fitness of an organism' w is not in fact a measurable property either of a population or of an individual, but a function of the *relative* fitnesses of *individuals*. Thus suppose that a population consisting wholly of As increases more rapidly than a population consisting wholly of Bs. It does not follow that $w_A > w_B$, or that As will replace Bs in a mixed population. For example, Bs may be cannibals and As not. If so, in a mixed population it may be that $w_B > w_A$, and Bs will replace As, and yet a population of As might increase more rapidly than a population of Bs. This objection is of more general application than the example of cannibalism might suggest. Characteristics which are favoured in intraspecific

competition often do not increase the probability that a species as a whole will survive.

(3) Even in a narrow sense, the theorem is not always true. For example, in a diploid sexually reproducing species, if at any locus the heterozygote is fitter than either homozygote, there will be an equilibrium at which there will be a genetic variance of fitness but no change in w with time. This last difficulty can be overcome by referring to 'additive genetic variance of fitness'. However, it does bring out the point that Fisher's theorem has no empirical content other than the laws of heredity; if certain assumptions about heredity do not hold, then the theorem does not hold.

Thus Fisher's theorem cannot help us to put an arrow on evolutionary time. Yet it is in some sense true that evolution has led from the simple to the complex: prokaryotes precede eukaryotes, single-celled precede many-celled organisms, taxes and kineses precede complex instinctive or learnt acts. I do not think that biology has at present anything very profound to say about this. If there is a 'law of increasing complexity', it refers not to single species, as does Fisher's theorem, but to the ecosystem as a whole. The complexity of the most complex species may increase, but not all species become more complex.

The obvious and uninteresting explanation of the evolution of increasing complexity is that the first organisms were necessarily simple, because the 'origin of life' is the origin, without natural selection, of entities capable subsequently of evolving by natural selection, and without selection there is no mechanism for generating a high degree of improbability—that is, complexity. And if the first organisms were simple, evolutionary change could only be in the direction of complexity.

Is there anything more interesting to be said? I do not know, but I have two comments to make. The first is that processes are known (for example, duplication) whereby the genetic material of an individual can increase. Even if the additional material is redundant or nonsensical, it does provide raw material for the evolution of increasing complexity. It is less easy to imagine processes leading to a loss of genetic material, since most losses will involve losses of functions essential for survival. The only exception is in the evolution of organisms (for example, viruses) living in an environment more complex than themselves, which may render previously essential functions unnecessary. It is significant that Spiegelman's 'evolving' RNA molecules[5] initially became simpler, and did so in an environment more complex than themselves.

The second comment is merely that we have at present no theory of evolving ecosystems, as opposed to the evolution of the species which compose them. In the absence of such a theory, it is hardly surprising that we can say little about the evolution of increasing complexity.

Summary

Two problems are discussed. The first is whether there has been time for evolution by natural selection to have occurred. The concept of a 'protein space' is introduced, and it is shown that a fundamental inequality, concerning the proportion of all amino acid sequences which form functional proteins, must be satisfied if evolution is to occur. The second is whether there is any biological law (analogous to the second law of thermodynamics) which enables us to put a time arrow on evolutionary processes. It is argued that Fisher's 'fundamental theorem of natural selection' does not meet this need.

NOTES AND REFERENCES

1 P. S. Moorhead & M. M. Kaplan (1967) *Mathematical Challenges to the Neo-Darwinian Interpretation of Evolution*. Philadelphia: Wistar Institute Press.

2 J. Maynard Smith (1961) The limitations of molecular evolution. *The Scientist Speculates*. (ed. I. J. Good) London: Heinemann.

3 For a review see J. L. King and T. H. Jukes (1969) Non-Darwinian evolution. *Science*, 164, 788–98.

4 R. A. Fisher (1930) *The Genetical Theory of Natural Selection*. London: Oxford University Press.

5 S. Spiegelmann (1968) The mechanism of RNA replication. *Cold Spring Harb. Symp. quant. Biol.*, 33, 101–24.

The Causes of Polymorphism

In this paper I shall try to give a general picture of the causes of genetic variability in natural populations from the point of view of a population geneticist. I will introduce algebra only when it is very simple and when it seems likely to be helpful to non-mathematicians. Mostly what I have to say is not new. However, the ideas on genetic variability in populations, like that of man, which have increased rapidly in numbers in the recent past, have not previously been published and indeed are not yet fully worked out.

From the beginning of population genetics there have been two schools of thought. On one hand, followers of R. A. Fisher have attempted to account for all changes and all variation in selective terms; on the other, followers of Sewall Wright have argued that chance effects arising because many actual breeding populations are relatively small are also important. Both views have been represented at this symposium. Those who like myself are students of J. B. S. Haldane pride ourselves on our ability to see both sides of the question.

This classic argument has broken out with renewed vigour as a result of the discovery of the vast amount of variation which can be shown to exist in proteins, mainly by electrophoresis. This variability, which is greater than population geneticists had expected, can be interpreted in two ways. On the one hand, it can be taken to show that selective mechanisms capable of maintaining polymorphism are more widespread than anyone had supposed. On the other, it has been argued that most of the protein variation which is found is selectively neutral, and is able to persist precisely because there are no selective forces acting on it. Along with the view that most protein variation is selectively neutral goes the view that most amino acid substitutions which have occurred during evolution were likewise selectively neutral, and were established by genetic drift.

Before pursuing the theory of selective neutrality further, it is worth considering briefly the selective mechanisms which have been suggested as capable of maintaining a genetic polymorphism.

Transient polymorphism

It may be that polymorphism observed in a natural population is not stable, and that the population has been caught in a moment of transition from one common allele at a locus to another. Although most species may be in a state of transient polymorphism at some loci, it is most unlikely that a majority of the protein polymorphisms recently described fall into this category.

Heterosis

If at any locus the heterozygote *Aa* is fitter than either homozygote *AA* or *aa*, it can easily be shown that a stable equilibrium exists, with both alleles common.[1] A classic case of heterosis at a single locus is that for sickle-cell haemoglobin,[2] since the heterozygote dies neither of anaemia nor of malaria. Heterosis for chromosomal inversions in *Drosophila pseudoobscura* has been fully investigated by Dobzhansky.[3] Whenever a polymorphism is long lasting in a relatively small population in a stable and uniform environment, it is natural to suspect heterosis.

Spatially varied environment plus migration

If in one part of the range of a species the homozygote *AA* is fitter and in another part *aa* is fitter, and if a degree of migration occurs, then there is likely to be a cline of gene frequencies, and local populations will be found to be polymorphic.

Frequency-dependent selection

If for any reason each of two homozygotes *AA* and *aa* at a locus is the fitter when it is rare, then there will be a stable polymorphism at some intermediate frequency. A number of ecological situations may give rise to such equilibria.

Predation. Predators have a frequency-dependent effect in either of two ways. Ford[4] pointed out that in Batesian mimicry there may be a balance between the frequencies of the mimic and cryptic forms of the mimicking species, because the mimic must not be too common relative to its models if predators are to leave it alone. Clarke[5] has emphasized that this formation of 'search images' by predators could give an advantage to the rarer morph of a species, even if it were more conspicuous, because individual predators would be more likely to form a search image of the commoner morph.

Disease. Haldane[6] discussed the possibility that some biochemical variants of a species might be favoured because parasites will tend to be better adapted to attack the common variants; by the same argument, rare variants of the parasite may be favoured because the host species is less likely to evolve resistance to them.

Selection in a varied environment. Maynard Smith[7] showed that if two ecological niches are available to a species, and if two genotypes exist such that AA is fitter in one niche and aa in the other, then there can be a stable polymorphism even if one allele is fully dominant in both niches. The equilibrium requires that the population density be separately regulated in the two environments, and that the selection pressure be high. This is a case of frequency-dependent selection for the following reason. When a particular genotype is rare, it finds few efficient competitors in its favoured niche, and so has a higher chance of surviving.

Cyclical selection

Haldane and Jayakar[8] investigated the possibility of a stable equilibrium between two alleles A and a if different homozygotes are favoured in different generations. They showed that such equilibria are possible, but are likely to exist only if selective pressures are large. Those interested in changes in gene frequency during cycles of population abundance should refer to this paper.

Other mechanisms maintaining stable polymorphisms are known; one is gametic selection favouring a gene which would otherwise be eliminated, as in the *t*-alleles in the house mouse.[9,10,11] A major difficulty in analyzing particular cases of polymorphism arises from the phenomena of linkage: it can be difficult to decide whether the stability arises from selection acting on the locus whose phenotypic effects are being observed or on some other locus closely linked to it.

Most participants in this symposium would probably attempt to interpret any case of polymorphism in terms of one of the above mechanisms, rather than as a result of neutral mutation or non-Darwinian evolution—or whatever phrase it is now fashionable to call what Wright would have called genetic drift. This may merely reflect the fact that this symposium is being held in the country of Fisher and not that of Wright. Whatever the reason, it may help to redress the balance if I next present what seems to be the strongest argument in favour of the neutral mutation theory; this is that the neutral mutation theory predicts that the rate of evolution for any class of proteins should be constant, and that there is some evidence for such uniformity.

The rate of non-Darwinian evolution

First, what is the evidence in favour of a uniform rate of evolution of proteins? The evidence is strongest for haemoglobin.[12] Figure 1 shows a phylogenetic tree of haemoglobin molecules, with an approximate time scale. The lines in this diagram represent descent by DNA replication, the polypeptide chains whose sequences have been determined being direct translations of these DNA molecules. Most bifurcations in the

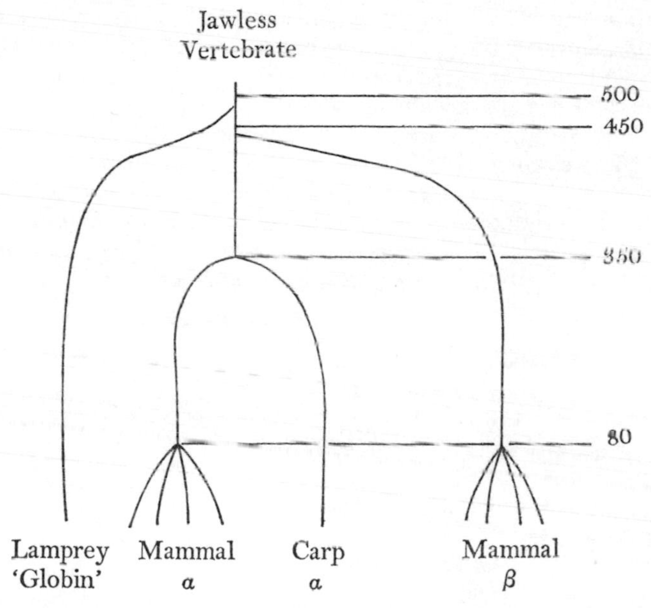

Figure 1. The phylogeny of haemoglobins. The figures on the right are approximate times in millions of years

diagram represent (as is customary in phylogenetic trees) a splitting of a single ancestral species into two; one of them represents the duplication, in an ancestor of the jawed vertebrates, of the gene which specified a single globin molecule into two initially identical genes whose descendants now specify the α and β chains of modern haemoglobins.

The amino acid sequences of existing polypeptide chains have been determined. By appropriate comparisons, it is possible to calculate the number of amino acid substitutions necessary to convert one into another. By dividing this number by twice the time to a common ancestor, one obtains an estimate of evolutionary rate in terms of amino acid substitutions per polypeptide chain per year. A further division by the number of amino acids per chain gives the rate in substitutions / site / year. A number of such estimates, based on different comparisons, are given in table 1. Some of these estimates are completely independent of one another, and others mainly so. The similarity of the estimates is striking.

The argument from the neutral mutation theory predicting this uniformity is so simple that it is worth giving.[13] The basis of the argument is the assumption that the majority of these substitutions are selectively neutral. Suppose that the rate of neutral mutation per generation per gene (that is, cistron) is u, and is constant for a given class of proteins. The idea is that most mutations will be harmful and will be eliminated. An occasional mutation is selectively neutral, and the fraction of mutations which are neutral is constant for a given type of protein; it will differ for different proteins, depending on the

Table 1. Average rates of amino acid substitution in the evolution of haemoglobins

Comparison			Substitutions / site / year $\times 10^{10}$
Human β	vs.	Lamprey Globin	12·8
Human β	vs.	Human α	8·9
Human β	vs.	Other Mammal βs	11·9
Mouse β	vs.	Other Mammal βs	14·0
Human α	vs.	Carp α	8·9
Human α	vs.	Other Mammal αs	8·8

variety and stringency of the selective requirements the protein must satisfy. Very rarely, a favourable mutation may occur and be established; these cases, though important in functional evolution, will be too rare to influence estimates of the rate of amino acid substitution.

Suppose that in any generation the population size is N. There are then $2N$ genes in this population, and $2Nu$ newly arising neutral mutations. If now we travel far enough into the future, we shall find that all genes in the population are ultimately derived from a single one of the $2N$ genes. Hence each gene has a probability $1/2N$ of ultimately being established; since a newly arising mutant makes the same contribution to fitness as any other, it too has a probability $1/2N$ of being established. Hence the number of new genes which arise and which ultimately are established per generation is $2Nu \times 1/2N = u$, a constant.

One complication is that the observational evidence suggests uniformity per year, not per generation as suggested by the preceding argument. This difficulty will disappear if it turns out, as may well be the case, that the mutation rate per generation in different mammals is proportional to generation time.

Evidence for uniformity of evolutionary rate for proteins other than haemoglobin has been reviewed by King and Jukes.[14] For some proteins the rule breaks down. For example, insulin has changed very little in most evolutionary lines, but is very variable among hystricomorph rodents. This latter fact could be explained on the neutral mutation theory only if it turns out that the selective restraints on insulin are less severe in hystricomorphs; I know of no reason why they should be.

Before leaving this topic, it will be useful later on to estimate for haemoglobin the value of u_e, the rate of neutral mutations per generation which are also electrophoretically recognizable. If changes at the nucleotide level were random, then about one third of all amino acid substitutions would involve a change in charge. An examination of the changes which have actually occurred in the evolution of haemoglobin shows that approximately the expected proportion of one third were charge changes. It is therefore justifiable, when testing the neutral mutation theory, to assume that electrophoretically recognizable changes are typical of all changes, at least for haemoglobin.

From table 1, the rate of substitution per site per year is approximately 10^{-9}. The number of sites per chain is 140, and a human generation lasts approximately 25 years. Hence the approximate neutral mutation rate is:

$$u_e = \frac{1}{3} \times 10^{-9} \times 25 \times 140 \simeq 10^{-6} \text{ per cistron per generation.}$$

Evidence from field studies

It is natural to attempt to use field data on protein variability to settle the argument between the proponents of selection and of drift. Some of the ways in which this might be done are as follows.

Measurements of selection

A popular way of demonstrating heterosis as a mechanism maintaining polymorphism is to show an excess of heterozygotes compared to expectation on the Hardy-Weinberg ratio. This was done[15] for *Drosophila pseudoobscura* males carrying inversions. The major snag to this method is that the ratio rests on the assumption of random mating. It is probably rare for there to be mating preferences based on solely morph differences (but see Sheppard[16] for an example). Unfortunately, if a sample contains members of different small inbreeding groups, it will depart from the Hardy-Weinberg ratio, although in the opposite direction to that expected in the case of heterosis. Examples of this difficulty have been given at this symposium.[17, 18]

An alternative method of measuring selection in natural populations is to compare gene or morph frequencies in young and old individuals. This was one of the first methods used in the measurement of natural selection.[19] It has been applied to mammalian populations by Van Valen,[20] and at this symposium by Lush.[21]

Association between environmental and gene frequency changes

If the frequency of a particular allele is found to vary consistently with the environment, this strongly supports a selectionist interpretation. An example, quoted by Lush at this symposium, concerns a locus in highland and lowland breeds of sheep. In general, Selander considers that the data he has presented to this symposium fit this interpretation, whereas Rasmussen[22] considers that the data on *Peromyscus* show gene frequency changes not associated with corresponding environmental change.

Environmental variation may be temporal rather than

spatial. In *Drosophila*, which can manage several generations a year, Dobzhansky[3] has found annual cycles in gene frequency. A mammal must evolve a genotype which will work the whole year round, but may nevertheless show cyclical changes in gene frequency in time with fluctuations in population numbers. Evidence has been presented at this symposium by Chitty[23] to show that such cyclical changes in gene frequency occur; if it can be established that the changes are genuinely associated with the population cycle, then the differences cannot be selectively neutral. Of course, it does not follow that the changes in gene frequency are a necessary driving component of the population cycle itself. Before this aspect of Chitty's theory can be investigated seriously, it will be necessary to have a precise model of the theory, either simulated or analytical, which can be shown to lead to oscillations.

Variability and population size

It is a necessary consequence of the neutral mutation theory that small populations should be less variable genetically than large ones. This affords a powerful method of testing the theory. Unfortunately, there are difficulties, which are discussed in more detail in the next two sections of this paper.

Population size and polymorphism

It is much easier to work out the consequences of the neutral mutation theory than of its selectionist alternatives. Consequently the strategy I shall adopt is to assume the truth of the neutral mutation theory and to work out the consequences of this assumption. When these consequences do not agree with observation, then some explanation in terms of selection is called for.

In a finite population, alleles will continuously be eliminated by chance, and new ones will arise by mutation. It was shown by Kimura and Crow[24] that when equilibrium has been reached between mutation and elimination,

$$I = \frac{1}{1+4N_e u} \qquad (1)$$

where I = probability that, at a particular locus, an individual chosen at random will be a homozygote,

u = rate of neutral mutation per generation at that locus, and

N_e = effective population size.

We usually cannot estimate I if all possible alleles at a locus are treated as different, but often have a fair idea if electrophoretically recognizable variants only are taken into account. Suppose then that $u \simeq 10^{-6}$, as estimated above for electrophoretic variants of haemoglobin. Then if

$$N_e = 10^6,\ I = 0{\cdot}20, \text{ and if}$$
$$N_e = 10^4,\ I = 0{\cdot}06\quad.$$

There is therefore a large difference between the degree of polymorphism to be expected in populations of large and small effective size. This difference can be used to test the theory, provided that populations can be found in which there is sufficient gene flow for the above formula to hold, and that the populations have been in existence for long enough for the equilibrium to be reached.

The time taken to reach the equilibrium may be the most serious qualification. It can be shown that it approximates, in generations, to the effective population size N_e. If the human population had kept at its present size since the Cambrian, we should still not have approached the equilibrium given in eq. (1). A possible way out of this difficulty is discussed in the next section. The objection does not always prevent the use of eq. (1). For example, suppose that a population of mice of effective number 10^4 or less has been reproductively isolated for 10,000 years from a much more abundant and effectively interbreeding population; one would, according to the neutral mutation theory, expect the isolate to show a markedly lower degree of polymorphism.

Before considering in more detail the problem of fluctuating population size, I must say more about what is meant by 'effectively interbreeding population' in the previous paragraph. What is required is that there should be sufficient gene flow to ensure that eq. (1) can be used with N_e equal to total population number, and not to the number in individual demes. The degree of gene flow to ensure this is much less than is required to ensure a good fit to the Hardy-Weinberg ratio.

I have considered this problem[25] for the 'island' model, in which a species is divided into a number of demes between which some migration takes place. If migration does occur, an individual is equally likely to migrate to any other deme whether close or distant. For mammals, a 'stepping stone' model in which migration occurs only between neighbouring demes would be more plausible, and should, if possible, be analyzed.

For the island model, if u is the neutral mutation rate, r is the number of local populations, and m is the probability that an individual born in one deme will breed in another, then there are two possible states of affairs. If $m > ur$, there is sufficient gene flow to ensure that eq. (1) can be used with N_e corresponding to the number of individuals in the species as a whole. If $m < ur$, there will be insufficient gene flow to ensure genetic similarity between demes and almost all hybrids between populations would be heterozygous. Hence if Anderson's suggestion[17] that in the house mouse there is insufficient gene flow to ensure genetic similarity between demes is correct, then if at any locus neutral mutations are possible, almost all inter-deme hybrids would be heterozygotes for that locus. In a sense, this conclusion is obvious, since if species integrity were not dependent on gene flow, it would have to depend on selection.

Summarizing this section, the following conclusions can be drawn:

(a) Equation (1) can be used only if an effectively interbreeding population of size N_e has been isolated for at least N_e generations.

(b) If, as is usually the case, hybrids between demes include an appreciable proportion of homozygotes at any locus, then one of three interpretations are possible. First, no neutral mutations occur at that locus. Secondly, there is sufficient migration to make it appropriate to use eq. (1) with N_e corresponding to the population as a whole and not to individual demes; this interpretation is of course possible only if N_e is small enough for eq. (1) to give a reasonable value of I. Thirdly, the population has arisen in the recent past from a smaller and effectively interbreeding population.

(c) The conclusions under (b) rest in part on an analysis of the 'island' model of population structure. It would be desirable to analyze the 'stepping stone' model in a similar way.

Polymorphism in a population whose size is changing

The most serious difficulty with eq. (1) is the great length of time required for the equilibrium to be reached—that is, a number of generations of the same order as N_e. The human population has increased by a factor of perhaps 10^4 in the past

500 or so generations. What type of variability could one expect from the neutral mutation theory in such a population? In view of the mathematical complexity of the equilibrium case[26] it may seem foolhardy to tackle the transient one. However, an approximate but adequate picture can be obtained in the following way.

If the present population consists of N_0 individuals, there are $2N_0$ genes. Of these genes, what fraction are copies, without further mutational change, of genes arising exactly n generations ago?

If N_n was the population size n generations ago, the number of new mutant genes arising then was $2N_n u$, where as before u is the neutral mutation rate per generation.

For each gene, (including the new mutants) present n generations ago the expected number of copies is now N_0/N_n. (This ignores mutations occurring in the last $(n-1)$ generations. This is permissible provided that $nu \ll 1$.) Hence the number of copies now of genes arising n generations ago is $2N_n u(N_0/N_n) = 2N_0 u$.

Hence the fraction of genes now which arose in any past generation is u. If $F_{n,m}$ is the expected fraction of genes now which originated between m and n generations ago, we have

$$F_{n,m} = (n-m)u \quad . \qquad (2)$$

This equation holds only if $nu \ll 1$; in the case of man, the equation can safely be applied over periods of up to one million years.

This equation will now be applied to electrophoretically distinguishable variants of haemoglobin in man. For such variants, it was estimated above that $u = 10^{-6}$ per generation. Hence approximately 4 genes in every 10,000 originated in the last 400 generations, or 10,000 years; that is, since the invention of agriculture and the subsequent increase in human numbers. Each individual allele would be very rare. Such rare variants of haemoglobin are known to exist[27] in about the right frequency. Their presence does not support the neutral mutation theory, since much the same result would be expected if there were a similar mutation rate to slightly deleterious alleles, since the probability of survival of a mutant for the first few hundred generations does not depend very critically on whether it is neutral or slightly harmful.

Consider now genes originating between 10^4 and 10^6 years ago, a period of about 50,000 generations. According to eq. (2)

approximately 5 per cent of existing genes should have originated during that period. A further 5 per cent should have originated between one and two million years ago, and so on. Now it seems clear that no such electrophoretically recognizable variants of haemoglobin exist. There are some common electrophoretic variants but they appear to be maintained in the population by heterosis, because of the resistance of the heterozygotes to malaria.

At first sight, the absence of such variants seems to be the death blow of the neutral mutation theory, at least as far as haemoglobin is concerned. Unfortunately, there is another possible explanation for their absence. $F_{n,m}$ is the expected fraction of genes arising between m and n generations ago. Hence there are two possibilities: either (a) 5 per cent of existing haemoglobin genes arose between 10^4 and 10^6 years ago; or (b) there is a 5 per cent chance that *all* existing haemoglobin genes arose during that period.

The latter possibility arises if there has been, during the past million years, a bottleneck in human numbers sufficiently small to give a reasonable chance that the haemoglobin locus became homozygous.

It is possible to give some idea of how small the effective population size would have to be to account for the absence of variability. Using the approach given by Kimura[26] and taking $u = 10^{-6}$, it can be calculated that N_e must fall below 10^5 before there is any approach to homogeneity. For $N_e = 10^4$ or less the population would ultimately have a high probability of being genetically homogeneous at a given locus. Using the 'diffusion' method developed by Kimura,[28] it can be shown that after a number of generations equal to approximately $2N_e$ an initially heterogeneous population would have an evens chance of becoming homogeneous.

Thus if the neutral mutation theory is true, the haemoglobin data require us to suppose that during the past million years there has been a 'bottleneck' in human numbers. This bottleneck could have been a single pair for one generation, or an effective population of 10,000 for half a million years, or something in between. The alternative is to conclude that there are no electrophoretically recognizable mutants of haemoglobin which are selectively neutral.

As on a previous occasion, we can choose between Darwin and the Garden of Eden.

Summary

Electrophoretic studies of proteins have shown that a large proportion of loci in natural populations are polymorphic. There are two possible explanations for these polymorphisms: (1) They are maintained by a balance of selective forces. The possible types of selective balance are reviewed. (2) The majority of polymorphisms are selectively neutral. If this 'neutral mutation' theory is true, then, at equilibrium between selection and mutation, there should be a higher proportion of polymorphic loci in large populations than small ones. Unfortunately, this prediction cannot easily be used to test the theory because the approach to equilibrium is very slow. An attempt is therefore made to derive the gene frequency distribution expected on the neutral mutation theory for a population which has recently increased in size. The method is applied to haemoglobin variants in man, and it is concluded either that few or no neutral mutations occur at the haemoglobin loci, or that human numbers passed through a bottleneck during the past million years.

References

1 R. A. Fisher (1930) *The Genetical Theory of Natural Selection*. Oxford: Clarendon Press.

2 A. C. Allison (1956) The sickle-cell and haemoglobin C genes in some African populations. *Ann. hum. Genet.*, 21, 67–89.

3 T. Dobzhansky (1951) *Genetics and the Origin of Species*. New York: Columbia University Press.

4 E. B. Ford (1953) The genetics of polymorphism in the Lepidoptera. *Adv. Genet.*, 5, 43–87.

5 B. Clarke (1962) Balanced polymorphism and the diversity of sympatric species. *Taxonomy and Geography*, pp. 47–70 (ed. D. Nichols). London: Systematics Association.

6 J. B. S. Haldane (1949) Disease and evolution. *Ricerca scient.*, Suppl. 19, 68–76.

7 J. Maynard Smith (1962) Disruptive selection, polymorphism and sympatric speciation. *Nature*, 195, 60–2.

8 J. B. S. Haldane & S. D. Jayakar (1963) Polymorphism due to selection of varying direction. *J. Genet.*, 58, 237–42.

9 D. Bruck (1957) Male segregation ratio advantage as a factor in maintaining lethal alleles in wild populations of house mice. *Proc. natn Acad. Sci. USA*, 43, 152–8.

10 R. C. Lewontin and L. C. Dunn (1960) The evolutionary dynamics of a polymorphism in the house mouse. *Genetics*, 45, 705–22.

11 P. K. Anderson (1964) Lethal alleles in *Mus musculus*: local distribution and evidence for isolation of demes. *Science*, 145, 177–8.

12 M. Kimura (1969) The rate of molecular evolution considered from the standpoint of population genetics. *Proc. natn Acad. Sci. USA*, 63, 1181–8.

13 M. Kimura (1968) Evolutionary rate at the molecular level. *Nature*, 217, 624–6.

14 J. L. King and T. H. Jukes (1969) Non-Darwinian evolution. *Science*, 164, 788–98.

15 T. Dobzhansky & H. Levene (1948) Genetics of natural populations. XVII. Proofs of operation of natural selection in wild populations of *Drosophila pseudoobscura*. *Genetics*, 33, 537–47.

16 P. M. Sheppard (1952) A note on non-random mating in the moth *Panaxia dominula* (L.). *Heredity*, 6, 239–41.

17 P. K. Anderson (1970) Ecological structure and gene flow in small mammals. *Symp. zool. Soc. Lond.*, No. 26, 299–325.

18 R. K. Selander (1970) Biochemical polymorphism in populations of the House mouse and Old-field mouse. *Symp. zool. Soc. Lond.*, No. 26, 73–91.

19 W. F. R. Weldon (1901) A first study of natural selection in *Clausilia laminata* (Montagu). *Biometrika*, 1, 109–24.

20 L. van Valen (1965) Selection in natural populations. III. Measurement and estimation. *Evolution, Lancaster, Pa*, 19, 514–28.

21 I. E. Lush (1970) The extent of biochemical variation in mammalian populations. *Symp. zool. Soc. Lond.*, No. 26, 43–71.

22 D. I. Rasmussen (1970) Biochemical polymorphisms and genetic structure in populations of *Peromyscus*. *Symp. zool. Soc. Lond.*, No. 26, 335–49.

23 D. Chitty (1970) Variation and population density. *Symp. zool. Soc. Lond.*, No. 26, 327–33.

24 M. Kimura & J. F. Crow (1964) The number of alleles that can be maintained in a finite population. *Genetics*, 49, 725–38.

25 J. Maynard Smith (1970) Population size, polymorphism and the rate of non-Darwinian evolution. *Am. Nat.*, **104**, 231–7.

26 M. Kimura (1968) Genetic variability maintained in a finite population due to mutational production of neutral and nearly neutral isoalleles. *Genet. Res.*, **11**, 247–69.

27 K. Sick, D. Beale, D. Irvine, H. Lehmann, P. T. Goodall, & S. MacDougall (1967) Haemoglobin $G_{Copenhagen}$ and Haemoglobin $J_{Cambridge}$. Two new β-chain variants of Haemoglobin A. *Biochim. biophys. Acta*, **140**, 231–42.

28 M. Kimura (1964) *Diffusion Models in Population Genetics*. London: Methuen.

The Origin and Maintenance of Sex

At the cellular level, sex is the opposite of reproduction; in reproduction one cell divides into two, whereas it is the essence of the sexual process that two cells should fuse to form one. In this essay I shall ask what selective forces were responsible for the origin of the sexual process, and by what selective process is it maintained. It is easier to ask these questions than to answer them; the fact that we cannot answer them with confidence is a challenge to evolution theory.

I was led to think about these questions after being involved in a controversy with Professor Wynne-Edwards on a quite different problem. It is Wynne-Edwards' thesis that animal population numbers are regulated by behavioural mechanisms which have evolved because they prevent the population from outrunning its food supply. Such mechanisms may involve individuals in refraining from breeding; they can therefore hardly confer a selective advantage on the individual, although they may confer an advantage on the group to which it belongs. Wynne-Edwards therefore believes that these mechanisms have evolved by a process of 'group selection', whereby gene frequencies change because some groups of related individuals are more likely to survive than others.

There are formidable difficulties for a population geneticist in any such explanation. This can be seen most easily from the following argument. Suppose that an individually harmful mutation occurs. This mutation can be eliminated from the population by a single selective death—that is, by the death of the first individual to carry the mutation. In contrast, suppose a mutation occurs that is beneficial to the individual but harmful to the group. The mutation will spread to all members of the group, and can be eliminated only by the elimination of the whole group. Thus the maintenance of a characteristic favourable only to the group requires N times as many selective deaths as the maintenance of an individually favourable characteristic, where N is the number of individuals in a reproductively isolated group. If groups are large, the selective cost of maintaining an 'altruistic' character will be prohibitive. It is therefore reasonable to attempt to explain the behavioural mechanisms described by Wynne-Edwards by selection acting at the level of the individual, and I think this can often be done.

There is however one property, that of sexual reproduction, which is almost universal, and for which the generally accepted explanation involves, implicitly or explicitly, a process of group selection. In its least precise form, this explanation states that sexual reproduction confers on a species a greater capacity for rapid evolutionary change, and consequently that when the environment changes, those species which reproduce sexually are more likely to survive. I do not doubt that this explanation is in some sense correct, but it raises more problems than it solves. In particular, if the advantages conferred by sex are long-term ones, conferred on a group as a whole, how could the complex genetic basis for sexual reproduction arise in the first place? And if the disadvantages of sexual reproduction, at least in multicellular bisexual organisms, are as great as they appear to be at first sight, why is not sexual reproduction more frequently lost?

Sex as an evolutionary advantage

The first step is to state more precisely why sexual species can evolve more rapidly. Evolution consists of changes in gene frequency. A gene frequency will not change under selection more rapidly in a sexual than in an asexual species; indeed, if the sexual species is diploid, some changes in gene frequency will occur much more slowly. Hence if only a single gene frequency is changing, sex is no advantage.

The advantages of sexual reproduction arise only when two or more genetic changes are being favoured simultaneously. This was recognized by Fisher, who concluded that 'the only groups in which we would expect sexual reproduction never to have been developed would be those, if such exist, of so simple a character that their genetic constitution consisted of a single gene'. However, it seems to me that Fisher does not specify precisely the circumstances in which sexual reproduction is an advantage. Thus he writes 'if . . . the mutation rates . . . are high enough to maintain any considerable genetic diversity, it will only be the best adapted genotype which can become the ancestor of future generations, and the beneficial mutations which occur will have only the minutest chance of not appearing in types of organism so inferior to some of their competitors, that their offspring will certainly be supplanted by those of the latter'. In other words, Fisher argues that in asexual species most beneficial mutations occur in individuals not destined to have descendants in the distant future, whereas in sexual species any beneficial mutation can be incorporated into the genotype of distant descendants, and that by virtue of this difference the rate of evolution in sexual species is more rapid.

I believe this argument to be fallacious, because, oddly enough, Fisher did not do the necessary sums. Thus suppose that a haploid population occupies an environment which changes, so that at two loci the initially common alleles, a and b, are at a selective disadvantage to the initially rare alleles A and B. Let P_{ab}, P_{Ab}, P_{aB}, and P_{AB} be the frequencies of the four genotypes. Evolutionary progress is measured by the rate of increase of P_{AB}, which is initially very small. It is shown in Appendix 1 that if initially $P_{ab} \cdot P_{AB} = P_{aB} \cdot P_{Ab}$, then this 'independence relation' will be maintained throughout the evolutionary change. Now the effect of sexual reproduction is to bring the genotype frequencies into agreement with the independence relation. If however the genotype frequencies already obey that relation, sexual reproduction will not accelerate evolution.

Now if in an asexual population the alleles A and B initially owe their presence to the recurrent mutation, $a \rightarrow A$ and $b \rightarrow B$, reaching an equilibrium between mutation and adverse selection, then it is easy to show that the 'independence relation' is satisfied. Thus in this simple case, in which according to Fisher's argument sexual reproduction ought to be an advantage, sex in fact makes no difference. Crow and Kimura have worked out the consequences of Fisher's argument quantitatively, and conclude that sexual species can evolve at rates many orders of magnitude greater than asexual ones. But they assume that favourable mutations are unique events, each type occurring once and once only. If this implausible assumption is dropped, their argument falls to the ground.

But suppose that there exist initially two environments, in one of which gene A is an advantage, and in the other of which gene B is an advantage. Populations will then evolve, one with P_{Ab} large and the other with P_{aB} large. Suppose now that a third environment becomes available for colonization by both populations (this could be a new area, or a transformation of one or both the existing environments). Then in the colonizing population $P_{Ab} \, . \, P_{aB} \gg P_{ab} \, . \, P_{AB}$, and sexual reproduction would enable P_{AB} to increase more rapidly than could happen in an asexual population.

In other words, if two genetically different populations are adapted to different environments, sexual reproduction makes possible the rapid evolution of a new population, carrying genes from both parental populations, and adapted to a third environment. Notice that the initial genetic adaptations were the result of natural selection. Thus another way of looking at the matter is to say that sexual reproduction makes it possible to utilize genetic variance generated by past natural selection to adapt rapidly to new circumstances. If existing genetic variance has been generated by mutation, as suggested in Fisher's argument, then sexual reproduction confers no advantage.

The origin of sex

It follows that sexual reproduction does confer a long-term advantage in enabling genes initially present in different individuals to be brought together in a single individual, but only if the 'species' (in this context, the group of individuals between which genetic recombination can take place) is divided into populations genetically adapted to different environments. Sex in this sense long preceded mitosis and meiosis; the processes of transduction and transformation achieve essentially the same end.

Sexual reproduction requires first that DNA from different ancestors be brought together in the same cell, and second that there be some mechanism of genetic recombination. The latter seems always to depend on a process of pairing between identical, or at least very similar, lengths of DNA, and on some process functionally equivalent to breakage and reunion of DNA molecules. It seems therefore that the enzymes required for genetic recombination could not have evolved because of the advantages conferred by sexual reproduction, because these advantages would not have existed until all the necessary enzymes had been perfected. In fact the same enzymes are probably used in repairing damaged DNA. As is so often the case in evolution, an organ—in this case a group of enzymes—which ultimately performs one function evolved in the first place because it performed another.

Thus genetic recombination may have been a by-product of selection for DNA-repairing enzymes. But in any case, before genetic recombination could occur, a means had to exist to bring DNA from different ancestors together in a single descendant. In higher organisms, the obvious selective advantage to such mechanisms arises from hybrid vigour; two homologous lengths of DNA may each be deficient, but in different cistrons, and may therefore complement one another. Unfortunately for this argument, bacteria do not often appear to utilize this particular advantage of diploids or heterokaryons. An alternative selective advantage for DNA transfer in bacteria has been suggested by Hayes. Organs such as *F* pili used in transferring DNA might in the first instance have developed under the instructions of viral DNA, which would thereby ensure its own transfer to a new bacterium. Only later would such organs be used to transfer bacterial DNA.

A sexual process involving a haploid-diploid cycle and meiosis, as found in eukaryotes, depended on the prior evolution of mitosis, and hence of centromeres, spindles, and centrioles. A theory of the origin of mitosis has been suggested by Sagan; what is important in the present context is that the relevant selective advantages were to the individual (or if Sagan is right, to the symbiotic pair, of which one provided the basal body of a flagellum, which evolved into both centromere and centriole) and not to the group. Once the machinery of mitosis had evolved, in organisms already possessing the enzymes needed for genetic recombination, the evolution of meiosis is not too difficult to understand. What is not clear is whether meiosis arose in organisms in which the main phase of the life cycle was haploid or diploid. It is possible that meiosis arose in an organism which could exist as a haploid, a heterokaryon, or a diploid. In the absence of meiosis the transformations open to such an organism would be:

haploid $\rightleftharpoons$ heterokaryon $\rightarrow$ diploid.

Heterokaryosis would evolve because of the advantages of hybrid vigour. But in heterokaryons there is no way of regulating the proportions of the two kinds of nuclei, and in cells with small numbers of nuclei there would be a constant danger of losing one or the other type. Diploidy, whereby the two sets of chromosomes of different ancestry share the same spindle, might therefore originate as a more stable way of propagating a particularly favourable heterozygous genotype. But it is still necessary to invoke the long-term evolutionary advantages of genetic recombination to explain the origin of meiosis.

The sex ratio

In eukaryotes with meiosis it is usual for there to be two sexes, and for each individual to have one parent of each sex. (These rules are not universal; for example, ciliates break the first and hymenoptera the second; the first rule is also broken in hermaphroditic animals and monoecious plants, which are discussed in detail later.) If these rules are obeyed, it is easy to show that natural selection will produce a sex ratio of unity. Thus suppose for example that there are more females than males. Then a male will have on the average more offspring than a female. Therefore a gene tending to cause individuals of either sex to have more male offspring, or tending to convert females into males, or to favour the survival of males at the expense of females, will increase under natural selection until the sex ratio is unity. The same argument applies in reverse if there are more males than females.

In microorganisms the situation is more complex, and the population genetics of 'sex ratio' is not understood. For example in *E. coli* there are $F-$, $F+$, and *Hfr* types, according to whether the F factor is absent, present in the cytoplasm, or incorporated into the chromosome. Transformations between the types can be represented as follows:

$$F- \rightleftharpoons F+ \rightleftharpoons Hfr \quad .$$

$F-$ are readily transferred into $F+$ by the transfer of an F factor, but the reverse transformation is rare and difficult to demonstrate. It is therefore puzzling that most *E. coli* outside laboratories are $F-$. The explanation is presumably that some $F-$ bacteria are at a significant selective advantage as colonizers of new habitats (if they enter a habitat already occupied by $F+$ bacteria they will be transformed). This in turn can be explained, since genetic recombinants are normally $F-$ bacteria which have received chromosomal material from *Hfr* bacteria. If this interpretation is correct, it is an interesting illustration of the advantages of sexual recombination.

In unicellular organisms, the disadvantages of sex are not great. The usual pattern is for asexual multiplication to be interrupted by sexual fusion only when conditions are severe and when continued multiplication would in any case be impossible. In view of this, and also of the fact that microorganisms commonly adapt to changed circumstances by evolutionary change as well as by individual physiological adaptation, it is not difficult to see why sexual processes, once evolved, should have been maintained.

But in multicellular organisms with separate male and female individuals the disadvantages of sex are severe. Suppose that in such a species, with equal numbers of males and females, a mutation occurs causing females to produce only parthenogenetic females like themselves. The number of eggs laid by a female, k, will not normally depend on whether she is parthenogenetic or not, but only on how much food she can accumulate over and above that needed to maintain herself. Similarly, the probability S that an egg will survive to breed will not normally depend on whether it is parthenogenetic. With these assumptions the changes shown in table 1 occur in one generation.

Table 1

		Adults	Eggs	Adults in next generation
parthenogenetic	♀♀	n ⟶	kn ⟶	Skn
sexual	♀♀	N	$\frac{1}{2}kN$	$\frac{1}{2}SkN$
	♂♂	N	$\frac{1}{2}kN$	$\frac{1}{2}SkN$

Hence in one generation the proportion of parthenogenetic females increases from $n/(2N+n)$ to $n/(N+n)$; when n is small, this is a doubling in each generation.

It follows that with these assumptions, the abandonment of sexual reproduction for parthenogenesis would have a large selective advantage in the short run.

It is well known that asexual varieties of plants arise quite commonly, and that their distribution, geographical and taxonomic, suggests that they are successful in the short term but in the long term doomed to extinction. Asexual varieties are much rarer among animals, although they do occur. It is not clear why this should be. Some possible reasons for the comparative rarity of asexual reproduction are:

(1) Meiotic parthenogenesis, followed by fusion of egg and polar body, or of the first two cleavage nuclei, is equivalent to close inbreeding. In naturally outbreeding species the decline in vigour caused by inbreeding might counterbalance the advantage of not wasting material on males. This argument does not apply to ameiotic parthenogenesis.

(2) In many mammals and birds, and some other animals, both parents help raise the young. In such cases parthenogenesis would usually be a disadvantage.

At first sight it seems that hermaphroditism, or monoecy in plants, eliminates the selective advantage of parthenogenesis. In a hermaphrodite species, no material is wasted on males, and no more resources need to be expended on sperm than are needed to fertilize the eggs produced. This argument I believe to be erroneous, at least in the case of hermaphrodites with external fertilization, for the following reasons.

In any species the number of eggs laid (or seeds produced) will be limited by some resource *R*. In a hermaphrodite the same individual will also produce sperm. It is reasonable to assume that the production of sperm will make demands on the same resource *R*, which must therefore be shared between eggs and sperm. The argument in the preceding paragraph amounts to saying that the major part of *R* will be devoted to eggs, only enough being devoted to sperm to ensure that the eggs are fertilized. This conclusion is an example of the use of what J.B.S. Haldane once referred to as "Pangloss' theorem"—that all is for the best in the best of all possible worlds. Unhappily, Pangloss' theorem is false. In this case it assumes that natural selection necessarily produces a result favourable to the species, regardless of selection at the individual level. This is not so, because individual selection is usually more effective than selection favouring one group or species at the expense of another.

In fact, it is shown in Appendix 2 that in hermaphrodites with external fertilization, or monoecious plants with compulsory cross-fertilization, the resource *R* will normally be divided equally between eggs and sperm, or ovules and pollen. In such cases hermaphrodites would on the average have only half as many surviving offspring as parthenogenetic females. However, the argument in the appendix does not apply to hermaphrodites with internal fertilization, or to self-fertilizing hermaphrodites, because in these cases individual selection will favour a limitation of the amount of sperm or pollen produced to that needed to ensure the fertilization of the available eggs. The conclusions to be drawn therefore vary according to whether a group has internal or external fertilization, as follows:

(1) In groups with external fertilization, hermaphroditism would not increase the reproductive potential of a species. It is the common mechanism of reproduction in land plants, presumably because it has the advantage that an individual can fall back on self-fertilization in the absence of near neighbours. It does not protect a species against the evolution of parthenogenesis.

(2) In groups with internal fertilization, hermaphroditism does increase the reproductive potential of a species. It may be for this reason that it has become the typical method of reproduction in plathyhelminthes and in gastropods. In the former of these groups it has proved to be a pre-adaptation to parasitism. It does protect a species against the evolution of parthenogenesis.

The argument in this section seems to lead to the conclusion that, except in the special cases of animals in which both parents care for the young, and of hermaphrodites with internal fertilization, metazoan animals would be expected to give rise frequently to parthenogenetic varieties. Since in fact this conclusion is false, the argument must leave something out of account. Ultimately what is left out of account is the long-term evolutionary advantage of sex. But the rarity of parthenogenetic varieties of animals suggests that this long-term selection acts, not by eliminating parthenogenetic varieties when they arise, but by favouring genetic and developmental mechanisms which cannot readily mutate to give a parthenogenetic variety. It is not clear how this has been achieved.

Appendix 1

The rate of evolution in sexual and asexual species

Consider first the evolution of an asexual haploid population varying at two loci, as follows:

genotype	ab	Ab	aB	AB
fitness	1	$1+K$	$1+k$	$(1+K)(1+k)$
frequency (generation n)	P_{ab}	P_{Ab}	P_{aB}	P_{AB}

Then if P'_{ab} etc. are the frequencies in generation $(n+1)$:

$$P'_{ab} = P_{ab}/T$$
$$P'_{Ab} = P_{Ab}(1+K)/T$$
$$P'_{aB} = P_{aB}(1+k)/T$$
$$P'_{AB} = P_{AB}(1+K)(1+k)/T$$

where

$$T = 1 + KP_{Ab} + kP_{aB} + (K+k+Kk)P_{AB} \quad .$$

Hence, if

$$P_{ab} \,.\, P_{AB} = P_{Ab} \,.\, P_{aB}$$

then

$$P'_{ab} \,.\, P'_{AB} = P'_{Ab} \,.\, P'_{aB} \quad .$$

Thus, if the 'independence relation' for genotype frequencies is satisfied initially, it will be satisfied in subsequent generations. The relevance of this fact is that all that is achieved by sexual reproduction is the production of a population obeying the independence relationship from one initially not satisfying it. It follows that sexual reproduction is an advantage only to populations which initially fail to satisfy the relationship.

I have discussed this problem in greater detail elsewhere.

Appendix 2
Resource allocation in hermaphrodites

Suppose that the number of eggs, n, and of sperm, N, produced by a hermaphroditic individual are limited by a common resource, R, of which an amount a is required to produce an egg and an amount b to produce a sperm. Then

$$an + bN = R, \qquad (1)$$

where R is a constant.

The argument is unaffected if some part of the quantities a and b are used in dispersing, protecting, or nourishing the gametes.

Let the typical members of a population at any time produce n_0 eggs and N_0 sperm. On the average, each typical member will have one surviving offspring as a female, and one as a male.

Now consider a mutant individual producing n' eggs and N' sperm. If each egg and sperm has the same chance of giving rise to a surviving offspring as those produced by a typical individual, then a mutant individual will produce n'/n_0 offspring as a female, and N'/N_0 offspring as a male. (This conclusion will not hold for a self-fertilizing hermaphrodite, or for hermaphrodites with internal fertilization, for which the fewer sperm are produced, the greater chance each sperm has of fertilizing an egg.)

Hence the total number of offspring produced by the mutant is

$$T = n'/n_0 + N'/N_0, \tag{2}$$

and substituting from eq. (1),

$$T = \frac{R - bN'}{R - bN_0} + \frac{N'}{N_0} .$$

Therefore

$$dT/dN' = \frac{1}{N_0} - \frac{1}{R/b - N_0} . \tag{3}$$

Hence if $N_0 < \frac{1}{2}R/b$, dT/dN' is positive; that is, mutations increasing N increase fitness, and selection will therefore increase N_0. Conversely, if $N_0 > \frac{1}{2}R/b$, selection will decrease N_0.

Thus there is a stable equilibrium when $N_0 = \frac{1}{2}R/b$, or when $bN_0 = an_0 = \frac{1}{2}R$. In other words, there is a stable equilibrium when the resource R is equally divided between eggs and sperm. The conclusion holds, however, only for hermaphrodites with external fertilization.